A Handbook To Australian

Seashells

On Seashores East to West and North to South

Barry Wilson

First published in Australia in 2002 by
Reed New Holland
an imprint of New Holland Publishers (Australia) Pty Ltd
Sydney

Reprinted in 2008, 2010, 2013, 2016, 2019, 2021, 2024

Level 1, 178 Fox Valley Road, Wahroonga, NSW 2076, Australia

National Library of Australia Cataloguing-in-Publication Data:

Wilson, B.R. (Barry Roberts), 1935–
A handbook to Australian seashells.

Bibliography
Includes index
ISBN 9781876334420

1. Shells—Australia—Identification—Handbooks, manuals, etc.
2. Mollusks—Australia. I. Title.

594.14770994

Publisher: Louise Egerton
Project Editor: Yani Silvana
Designer: Avil Makula
Illustrator: Edwina Riddell
Cartographer: Ian Faulkner
Production Controller: Wendy Hunt
Reproduction: Pica Digital, Singapore
Printer: Toppan Leefung Printing Ltd,China

Contents

The Shells 25

Acknowledgements

Many fine shells in the Bellview Shell Collection, Witchcliffe, southern Western Australia, were photographed for this book with the kind permission and assistance of Peter and Kath Ignoti. Peter Clarkson also made some beautiful specimens available. The help of these wonderful people is sincerely acknowledged. In addition, some specimens were photographed from the collections of the Western Australian Museum and the assistance of the Museum staff is also greatly appreciated. Mr Clay Bryce of the Western Australian Museum kindly provided several photographs of chitons from his own collection.

Introduction

Collecting seashells on the seashore is an activity that delights all ages. There is a mystery about the exquisite symmetry of these objects that sparks the imagination.

A collection of shells can be an aesthetic pleasure but it can also lead to an understanding of the natural history of one of the greatest groups of the animal kingdom, the molluscs (phylum: Mollusca). Seashells are created and have been inhabited at one time by living creatures belonging to this group. Each species manufactures a characteristic shell that is discernibly different from those of other species, creating a seemingly endless variety of form and colour. The design of the shell reflects the genetic heritage of the species and reveals much about the lifestyle of the animal that made it.

Australian shores are blessed with a greater variety of shells — tens of thousands of different kinds — than almost any other part of the world. This book introduces the collector and the student to some of the common shells found along the Australian coast. It is far from being comprehensive as complete coverage of all the many groups of shelled molluscs would require a far weightier work than this one. Here only the obvious and common species — those frequently cast ashore or that live along the shore in shallow water — have been chosen to represent the range of diversity of this great phylum. Many families are not represented at all due to lack of space.

The majority of living molluscs are small and have very tiny shells, but few collectors and students have the facilities to study the smaller shells. It is also technically difficult to illustrate them in a book like this. Consequently only species that have medium-sized or large shells, that is, larger than 10 mm, are included. Many molluscs do not have shells at all and these, with one exception, are also excluded from this work.

Molluscs

While seashells are fascinating in their own right, they are but artefacts built by living creatures, molluscs, whose lives are as varied and as interesting as the shells they make. They include bivalves (like pipis), gastropods (like snails), chitons (also known as coat-of-mail shells) and cephalopods (like nautiluses and octopuses). Most of these modest animals, lacking any hard internal skeleton, compensate by secreting a hard external casing made of calcium carbonate to protect their bodies and provide for muscle attachment and support for their soft parts.

Names and Classification of Molluscs

Because seashells are built and inhabited by living creatures, they are named and classified according to the rules used for all living things. To the uninitiated, the bodies of a chiton, a snail, an oyster and a squid may seem so wildly different that classifying them together in the same phylum is incredible, yet they all share the basic structure described under the heading The Molluscan Body on page 7. It is the variation on this theme that provides the basis for the classification within the group.

NAMES

Familiar plants and animals generally bear names assigned to them through popular usage. All species described in scientific terms, whether they be widely familiar or not, also have formal scientific names. In this book both common and scientific names are used.

Under the binomial system established centuries ago, each animal and plant species has two scientific names, a generic name and a species name, both of which are italicised. They are normally used together, the generic name first (capitalised) followed by the species name (not capitalised); e.g. the scientific name of the Tiger Cowry is *Cypraea tigris*. There is a set of international rules that governs the priority of scientific names.

There are no such rules for the use of common names. The same shells may have different names in different languages and even in different places using the same language. There are widely used common names for most families of molluscs, although there are inconsistencies. Common names for species are more difficult. Many species, though scientifically named, have never been given common names. In this book the common names that are used follow previous usage where possible. Where there has been no prior common name, one has been derived from the scientific name. Scientific species names are usually Latinised terms that describe the species, the locality where it may be found or a person after whom it was named. It is usually simple to anglicise them.

CLASSIFICATION

Seashells belong to the phylum Mollusca, which traditionally contains seven classes. Their classification is constantly changing as new information and new study methodologies are discovered. For our purposes, following the traditional classification, the classes are:

- **Aplacophorans** (class Aplacophora). These tiny worm-shaped marine creatures mostly live in deep water and are rarely seen, even by professional zoologists.

- **Chitons or coat-of-mail shells** (class Polyplacophora). There are several hundred species, all marine, including a number that are conspicuous residents of the intertidal zone on Australian rocky shores. Their characteristic shells have eight articulating parts (plates) and the animals have a very simple anatomy.

- **Monoplacophorans** (class Monoplacophora). These small, limpet-like molluscs are from the deep oceans. Of a handful of living species, none are recorded in Australian waters.

- **Tusk shells** (class Scaphopoda). Of several hundred species, only a few are found on Australian sandy shores. All are marine sand-dwellers with elephant tusk–shaped shells.

- **Gastropods** (class Gastropoda). Snails and slugs make up this, the largest of the classes, with at least 90 000 species worldwide. They live in marine, freshwater and terrestrial habitats. The shelled gastropods, also known as univalves because the shell consists of one piece, include such well-known creatures as garden snails, abalone, conchs and cowries. The majority of the marine gastropods are known as prosobranchs and they usually have a shell. There is also a very large group of slugs, known as opisthobranchs, only some of which bear shells. Air-breathing gastropods, including most of the land snails and slugs and a few intertidal marine species, are known as pulmonates.

- **Bivalves** (class Bivalvia). This is the second largest class with about 15 000 species living in marine and freshwater habitats. Perhaps the best known are the edible mussels and oysters, the pearl oysters and the giant clams, but a large variety of other bivalve shells, representing many species, may be found around Australian shores.

- **Cephalopods** (class Cephalopoda). This very diverse class has several thousand species, all marine, including nautiluses, squids and octopuses. Only a few of them are shelled.

Most shells found along Australian shores are those of prosobranch gastropods or bivalves and these classes comprise the majority of the species illustrated in this book. Only a few selected species of chitons and shelled cephalopods and opisthobranch gastropods are illustrated to represent those groups. For a complete classification of Australian marine shells, consult the major technical work entitled *Mollusca: the Southern Synthesis. Fauna of Australia*, Beesley et al. 1998, which can be found in most Australian libraries.

The Molluscan Body

The bodies of molluscs are unsegmented and soft, that is, without a skeleton. In most kinds the soft organs are grouped together in what is known as the visceral mass, atop a muscular 'foot' which is the principal locomotory organ.

A special feature of molluscs is the mantle, a thin, skin-like layer of tissue draped over the visceral mass, enclosing a space between it and the body. This space is called the mantle cavity and it contains certain important organs, including the gills and the osphradium, an organ used to 'smell' food, toxins and chemical reproductive messages in the water. The reproductive system, excretory system and alimentary system also open into the mantle cavity. Water is drawn into the space, bringing oxygen, and passed out again via the gills carrying away carbon dioxide and body wastes. Excretory wastes from the kidney aperture and alimentary wastes from the anus are carried away by the outgoing (excurrent) water as it leaves the mantle cavity. Eggs and sperm also may be shed from the reproductive aperture into this outgoing water.

Movement of all these materials through the mantle cavity is effected mainly by the beating of countless tiny, hair-like cilia, facilitated by mucus secreted by the mucous gland.

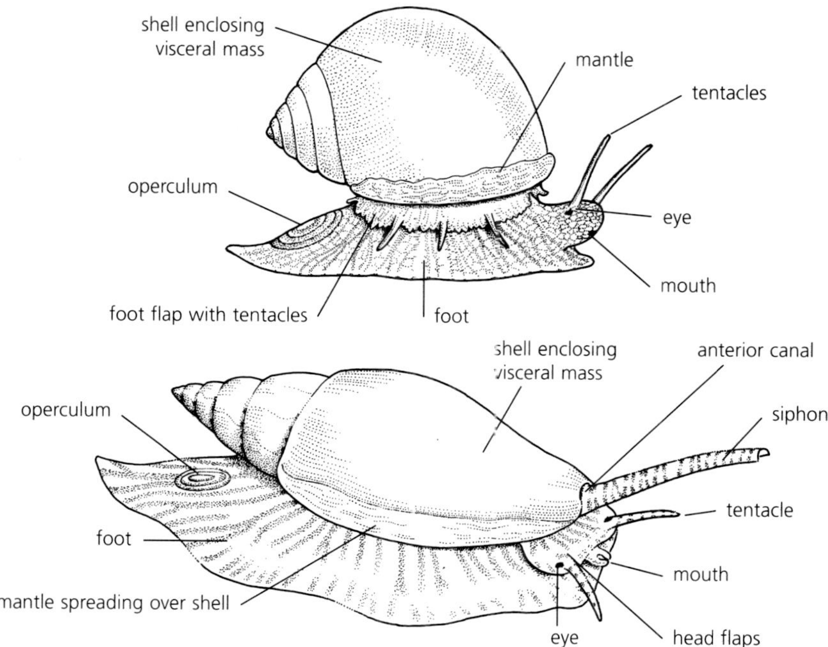

Top *Simple gastropod such as a trochid — no extensible siphon.*
Bottom *More complex predatory gastropod such as a volute — with an extensible siphon and a wide foot well adapted to digging.*

How the Shell Is Made

Calcium is taken into the body either with food or directly by absorption from the water. Within the body it combines with carbonate to form calcium carbonate (lime). The calcium carbonate that forms the shell, and the pigments that colour it, are secreted by glands in the mantle. The precision by which the material is laid in position, resulting in the unique form and sculpture of each species, is one of the great mysteries of life. As the calcium carbonate is deposited the mantle secretes pigments that are laid down as spots, lines or other marks, depending on the movement of the mantle.

Shell Growth

As the animal grows it must enlarge the shell to accommodate itself. It does this by adding shell material around the edges of the valves (bivalves) or lips of the aperture (gastropods).

In the case of tubular gastropods (all gastropods except limpets) the process of enlargement of shell volume is like adding bricks to the rim of a chimney to increase its height. But in addition to increasing the length, the diameter of the tube is also progressively increased, creating a funnel shape. The shell tube of most gastropods is coiled. To achieve this spiral growth, the new shell material is not added evenly around the rim. The rate of the increase in diameter and the position on the rim where the material is deposited are the primary factors determining the ultimate form of the shell. The complexity of the mathematics of this kind of growth and the consistency with which each species produces precisely the form that characterises its kind are beyond comprehension.

To thicken and strengthen the shell as its size increases the animal adds material to the inner surface. Many species also add ribs and buttresses that increase the strength of the structure, just as an engineer would do with a building.

Shell growth is often not continuous but periodic. In such cases strengthening ribs may be formed at the edge of the lip during the quiescent phases. Some molluscs do not have a natural maximum size but keep growing, albeit at a progressively slower rate, until they die. Others reach a maximum size early in adult life and further shell growth is limited to thickening. Some molluscs live only for a single breeding season while a few, like the giant clams and baler shells, may live for many decades. There are few data but it is likely that most molluscs will live for several years.

Shell Parts

There is a host of terms used to describe shells. Many of them are general terms whose definitions may be found in a good dictionary. Others are special terms used solely or mainly in conchology (the study of shells). In the language of conchologists some common terms are used in a special way: the most widely used of these are given in the Glossary on page 178.

GASTROPOD SHELLS

Gastropod shells are sometimes called univalves because they comprise one piece. Some are simple conical structures that are held over the body. This is the 'limpet' form that has been adopted by several groups unrelated to each other, most notably the true patellid (Patellidae) and acmaeid (Acmaiediae) limpets, the 'key-hole' limpets (Fissurellidae) and the air-breathing pulmonate limpets (Siphonariidae). Limpet shells are attached to the muscular foot by **adductor muscles**, enabling the animal to clamp down hard onto the rock or whatever hard surface it is living on, protecting the body beneath it. Key-hole limpets usually have a hole near the apex or a slit in the margin (the **fissure** from which these fissurellids take their name) through which the outgoing water currents are ejected from the mantle cavity.

The majority of shelled gastropods have coiled, tubular shells. The apex or point of the tube represents the larval shell (the **protoconch**) and as the shell grows the tube expands and coils around a central axis. Each coil is known as a **whorl**: the last one, that is, the largest one that contains most of the animal's body, is called the **body whorl**. Generally the coiling is right-handed — when the shell is viewed with the spire tip pointing directly towards you the coiling is clockwise. Right-handed shells are called dextral. The rare cases of left-handed shells are known as sinistral.

Usually the expanding coiled tube is drawn out, the early whorls forming a cone-shaped **spire**. These shells are measured by their height (or length) from the anterior end to the tip of the spire and their width across the widest part of the body whorl. Shells without a spire, like nautilus shells, are measured by their diameter.

The coiling is usually 'tight', that is, each whorl is tightly cemented to the side of the whorl before it, the line of contact being called the **suture**. Sometimes, if the coiling is not tight, there is a cavity up the central axis, which appears at the base of the shell as a hole known as the **umbilicus**.

The opening of the shell is called the **aperture** (or mouth). The inner side of the aperture is usually thickened with extra layers of shell material referred to as callus. When the coil is tight, thick callus may form a strong central pillar that runs through the central axis of the shell. The thickened part of the pillar exposed along the inner lip (inner edge of aperture) is called the **columella**, which may be simple, nodulose, lirate or spirally plaited.

The outside edge of the aperture is called the **outer lip**. It, too, may be thickened and bear marginal nodules that are usually referred to as 'teeth', though of course they have nothing to do with mouthparts or the ingestion of food. If the edge of the lip is smooth it is called **simple** but it may be **serrate** (toothed) or **crenulate** (notched). (These terms apply also to bivalves — see illustrations on page 13.)

The shape of the aperture is also an important character when describing gastropod shells. It may be round, elliptical or pear-shaped. If the margin of the aperture is uninterrupted it is said to be **entire**. But most shells have a u-shaped notch called the **anterior canal** in the anterior margin. It supports a fold in the mantle known as the **siphon**, which channels the water entering the mantle cavity. There may also be a notch in the other (posterior) end of the aperture margin supporting an 'anal' siphon that channels the outgoing water; this is called the **posterior canal**. Sometimes the margins of the anterior canal are extended to form an elongate tube, rather like a diver's snorkel, supporting a long siphon. More rarely there is an extended posterior or anal canal as well.

The outer surface of most gastropod shells is sculptured with either axial or spiral ribs or both. Fine ribs are usually called **cords** and thick ones **lirae**. When the ribs are high and thin they are called **lamellae**. In both gastropods and bivalves many terms are used to describe external sculpture, depending on the relative strength of the axial and spiral elements, such as nodulose, granose, sulcate, striate, imbricate, lamellate, cancellate. These terms are defined in the glossary.

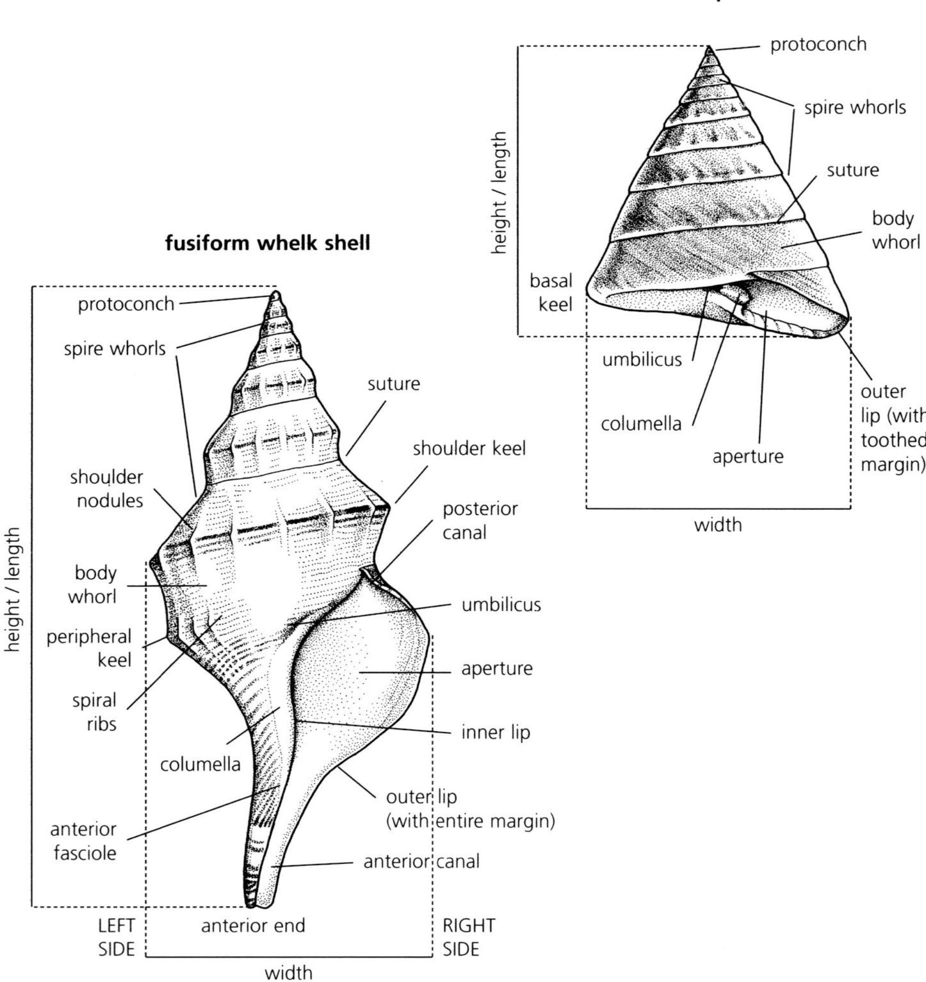

Two typical gastropod shells.

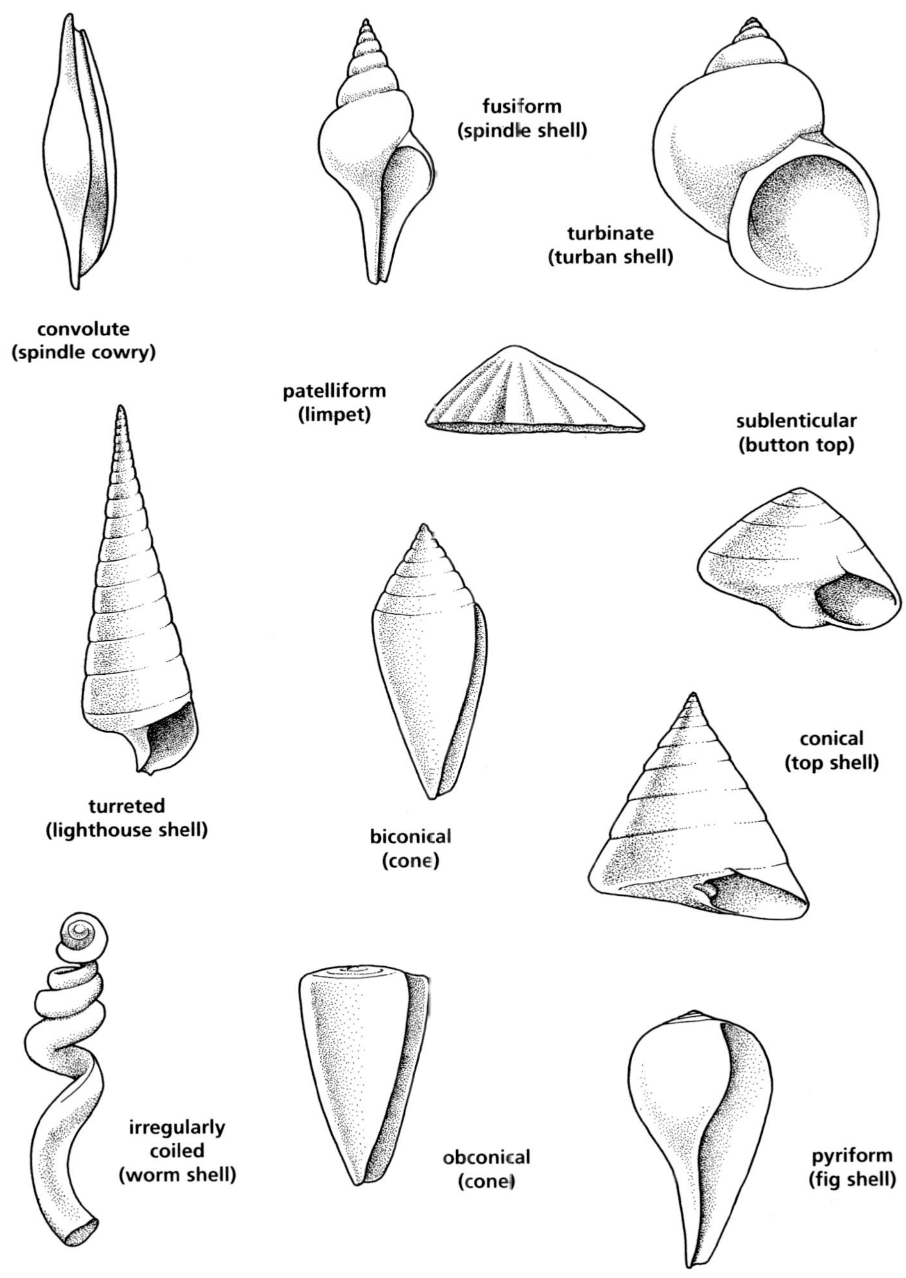

Terms describing the various shapes of gastropod shells (after Wilson & Gillett 1971).

BIVALVES

Bivalve shells consist of two parts called **valves**. The left and right valves are articulated together along the upper margin. Usually both valves are concave dishes which, when held together, enclose a space between them. The valves of the larval shell appear at the top as points that are called the **umbos**. If the umbos are central, so that the anterior and posterior sides of the shell are more-or-less equal, the shell is said to be **equilateral** (**inequilateral** if the umbos are not central). If the two valves are of equal size and have concavities of equal depth they are said to be **equivalve** (**inequivalve** if one valve is more concave than the other.)

The umbos of most bivalves point forwards. If you hold the bivalve shell with the umbos at the top and pointing away from you (that is, anteriorly) the left and right valves are easily identified (see venus clam opposite). When the umbos are central (see bittersweet clam opposite), it is very difficult to determine which valve is which.

Bivalves are measured by their height, length and width. Height and length are shown on the illustrations opposite. When the valves are more or less circular the size is expressed as diameter, because height and length are equal. Width is the distance across the two valves when they are held together. The measurement given in Part Two of this book is the greatest.

Size, shape and position of **muscle scars** on the inner surface of the valves are important characters in the classification and identification of bivalves. The valves are held together by adductor muscles that cross the space between the valves and attach to their opposite inner surfaces. There are usually two sets of adductors, an anterior set and a posterior set, each making muscle scars at the insertion positions. The posterior adductor muscles and their scars are often larger than the anterior ones. There are also muscles that retract the foot and **byssus** (a rope-like structure that some bivalves use to anchor themselves to a firm surface — see illustration **a** on page 17) and these are called the **pedal** and **byssal retractor muscles**. These also make scars where they attach to the inner surface of the valves. Finally, the edges of the mantle are attached around and just inside the perimeters of the valves. The mantle is attached to the shell by tiny muscles forming a distinct line around the inner side of the valves. This is known as the **pallial line**. In some families the pallial line has a posterior indentation representing the position of the siphons and this is called the **pallial sinus**.

Along the dorsal margin, below or behind the umbos, there is a stiff elastic structure called the **ligament** joining the valves. When the adductor muscles are contracted and the valves are shut together, the ligament is compressed; but when the muscles relax the ligament springs and holds them apart. The ligament may be narrow and long or short and stout.

Also on the dorsal margin of each valve is a swelling known as the **hinge**, that is, the fulcrum against which the two valves swing open and shut. The hinge has interlocking, tooth-like nodules that are called the **hinge teeth**. The number, shape and position of the hinge teeth are critical characters for the identification of bivalves. The two most common types of hinge teeth are the **taxodont** type (typified by the arc shells) consisting of a long series of rectangular nodules, and the **heterodont** type consisting of a few teeth fitting into sockets on the opposite valve. Most commonly the heterodont teeth include a cluster of peg-like **cardinal teeth** centred directly below the umbos and more elongate **lateral teeth** at the anterior and posterior ends of the hinge, further away from the umbos.

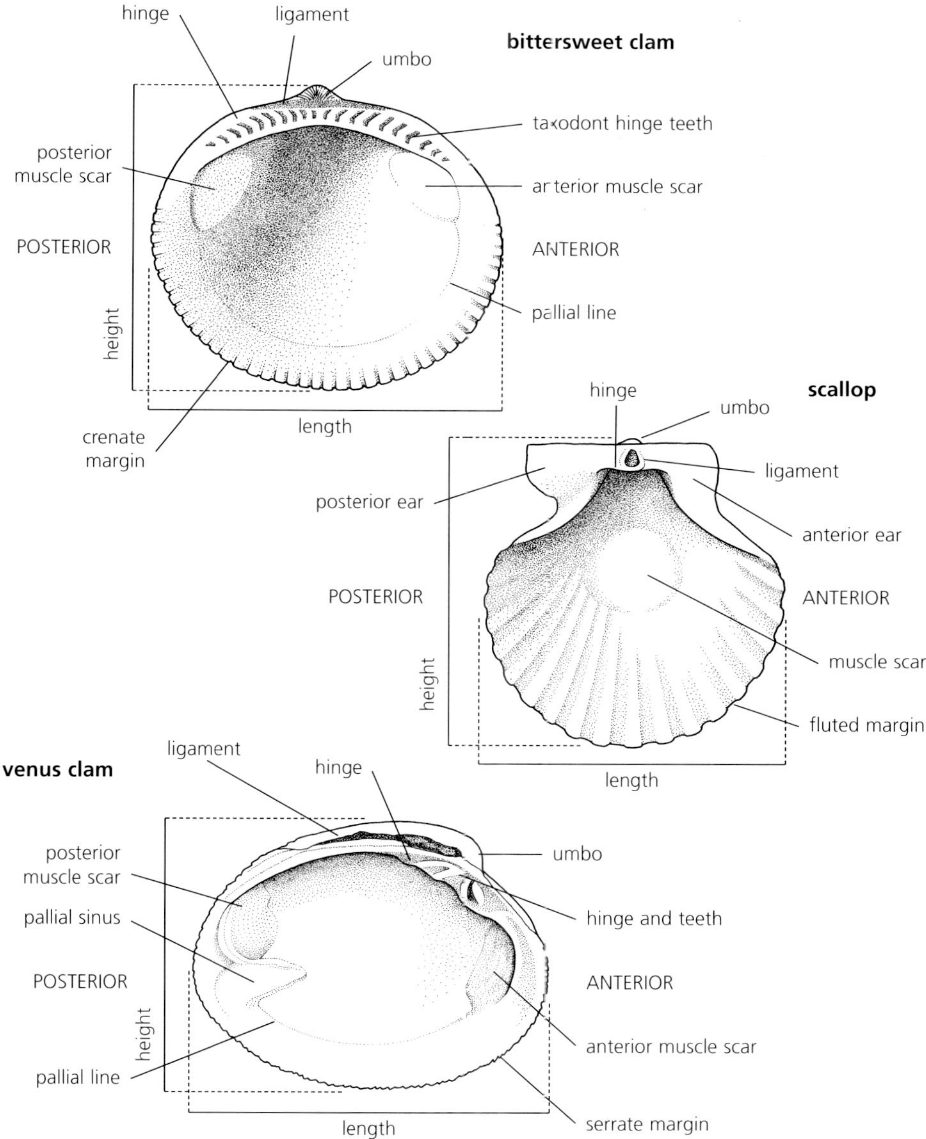

Three typical bivalve shells (all left valves shown from their inner sides):
bittersweet clam — equilateral with two muscle scars and taxodont hinge teeth
scallop — equilateral with a single muscle scar and no hinge teeth
venus clam — inequilateral with two muscle scars and heterodont hinge teeth.

Geographic Distribution

A shell collector from Bass Strait would find few familiar shells on the shores of northern Queensland or the Pilbara coast of Western Australia. Many species would seem vaguely familiar because of relationship at the generic or family level, but the species would be nearly all different. The marine animals and plants of the temperate waters of southern Australia have had a very different evolutionary history to those of the tropical north. The differences are so pronounced that two major biogeographic regions are recognised around the Australian coast — the temperate Southern Australian Region and the tropical Northern Australian Region.

Many of the species in the north are widespread throughout the tropical waters of the Indian and western Pacific oceans and the Northern Australian Region is regarded as a subdivision of the vast Indo-Pacific Faunal Region. In contrast, the majority of temperate species in the south are peculiar to southern Australian shores and the Southern Australian Region is regarded as a distinct region in its own right.

On the eastern (Pacific Ocean) and western (Indian Ocean) sides of the continent the tropical and temperate faunas overlap. On the west coast the Leeuwin Current, originating on the North West Shelf, brings tropical water and a load of tropical species' larvae far down into the temperate zone. On the east coast the East Australia Current produces a similar result.

This book cites the geographic distributions of species according to a general biogeographic classification of the Australian coast, which is diagrammatically represented on the map opposite. Note that a majority of northern Australian species also range through the Indo-Pacific region (see map opposite below), or at least the central part of it, and this is indicated in the text when it applies.

Collectors interested in exploring in detail the interesting subject of distribution patterns of Australian marine molluscs could begin by consulting the following report published by Environment Australia: *Interim Marine and Coastal Regionalisation for Australia — An ecosystem-based classification for marine and coastal environments*, IMCRA Technical Group, June 1998, eds R. Thackway and I. D. Cresswell.

Habitats

People collecting shells are most likely to pick up empty ones cast ashore after their makers have died. But marine molluscs live in a variety of habitats. Most are slow-moving, secretive creatures. An inexperienced observer may swim or walk over a coral reef, for example, and see very few molluscs where thousands exist. Many hide during the day, emerging to go about their business at night. Learning the habitats and habits of the many species and where and how to look for them is one of the joys of marine natural history.

Although a few marine molluscs (such as the winkles) live above high-tide level (supratidal zone) you are more likely to find living animals in the rich and diverse habitats between high- and low-tide levels (intertidal zone) when the tide is low, or by donning a face mask to explore the shallow water below low-tide level (subtidal zone).

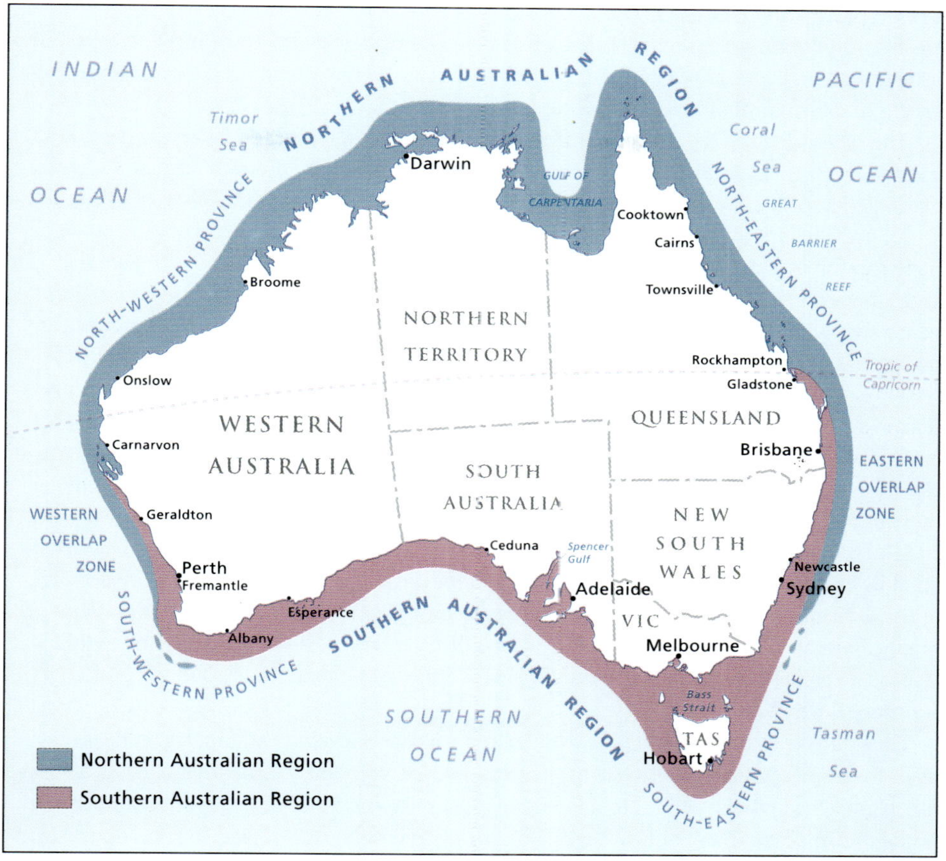

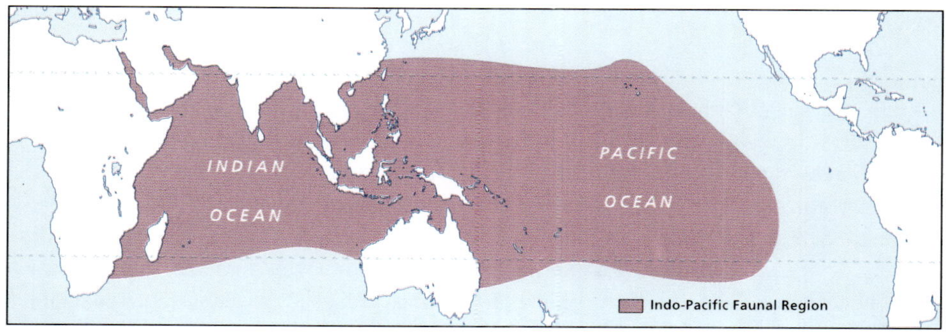

The two major biogeographic regions around Australia's coast.

Northern Australian Region

Southern Australian Region

The Indo-Pacific Faunal Region, to which the Northern Australian Region belongs.

THE INTERTIDAL ZONE

The intertidal zone, where terrestrial and marine ecosystems interact, is the most accessible zone to find molluscs. There, for brief periods each day, the sea retreats and you can examine marine creatures in their natural habitats.

The rise and fall of the tide has profound effects on life in the intertidal zone, where the dominant factor is periodic exposure to air, sun and drying winds. Molluscs have developed physiological or behavioural ways to deal with exposure in order to live there successfully.

On rocky shores, molluscs can limit exposure of their tissues by simply clamping down tightly to the rock or retreating to a rock pool or beneath the shelter of a ledge, where they can avoid the desiccating effects of the sun and wind. Some gastropods have an operculum (lid) which tightly closes the aperture — but of course, once the aperture is closed, the animal can no longer use the foot to cling to the rock.

On sandy or muddy shores in the intertidal zone most molluscs avoid exposure by burrowing or retreating to shallow pools. Many bivalves have developed snorkel-like siphons, one for water intake and one for water expulsion, which allow the animal to bring water with essential oxygen and suspended food into the mantle cavity. In this way these burrowing bivalves can live protected below the surface of the sediment (see illustration oppostite).

The problem of exposure is most acute in the upper part of the intertidal zone because animals living there are exposed most often and for the longest periods. The uppermost edge of the intertidal zone is briefly flooded by the sea only during periods of spring tides. Very few marine animals are able to survive there. At the other extreme, the very lowest part of the intertidal zone is rarely exposed to the air, that is, during low spring tides, and the period of exposure is very brief.

This difference in exposure time from extreme low-tide level to extreme high-tide level leads to distinct 'zonation patterns' in the communities of molluscs and other animals and plants inhabiting the intertidal zone. Certain species, especially those living in the upper part of the intertidal zone, can only live within narrow parts of the tidal range. They occupy distinct horizontal bands at different levels around the shore. There may be a band of mussels, a band of barnacles, a band of limpets and a band (in the supratidal zone) of winkles, one above the other. This sequence of bands from top to bottom of a shore profile is known as 'vertical zonation'.

Vertical zonation is most noticeable on rocky shores. Within a single biogeographic region of the coast, zonation patterns on rocky shores will be similar, with the same species occupying the same zones. Rocky shores of different biogreographic regions will exhibit similar zonation patterns but the species may be different. The diagrams on pages 18 and 19 illustrate this point with examples from the south-east and north-west Australian coasts.

Vertical zonation is much less pronounced on sandy and muddy shores, perhaps because the intertidal zone is spread over a wider horizontal distance. Nevertheless, the same principle applies. The upper intertidal zone of beaches is normally poor habitat for molluscs, but throughout the world species of the bivalve family Donacidae ('pipis') live there, usually in the coarse sand at the bottom of the beach slope that is swept by the breaking waves. There are seven species of this family on beaches around the Australian coast.

SUBSTRATE

The nature of the surface upon which the animal lives is another very important aspect of coastal habitats. Rocky shores alternate with sandy beaches and there are many areas with wide intertidal mud or sand flats.

The first challenge for dwellers on rocky shores, especially in wave-swept places, is how to secure their position. Chitons and many gastropods such as limpets and abalone inhabit reefs and rocky shores. They attach themselves to firm surfaces such as rocks by means of their flat-soled muscular foot. Byssal-attached bivalves are sessile creatures living on reefs and rocky shores, or at least attached to firm objects lying in soft sediments. Some sessile bivalves, like the rock oysters, have resorted to cementing one of the valves to a hard object. Others, like the date mussels, bore into rocks (especially limestone and coral rock). Other gastropods that need to maintain their mobility hide in crevices, under stones or among the fronds of attached algae.

Inhabitants of soft substrates (sand and mud) have different problems. Their challenge is to find protection within the sediment without being buried by it. Some soft-substrate gastropods, like dog whelks, olives and moon snails crawl about or bury themselves out of sight remarkably quickly. The hatchet-shaped or tentacle-like foot of bivalves is not suited to adhesion to hard surfaces but may be a useful implement for digging or crawling in sediments.

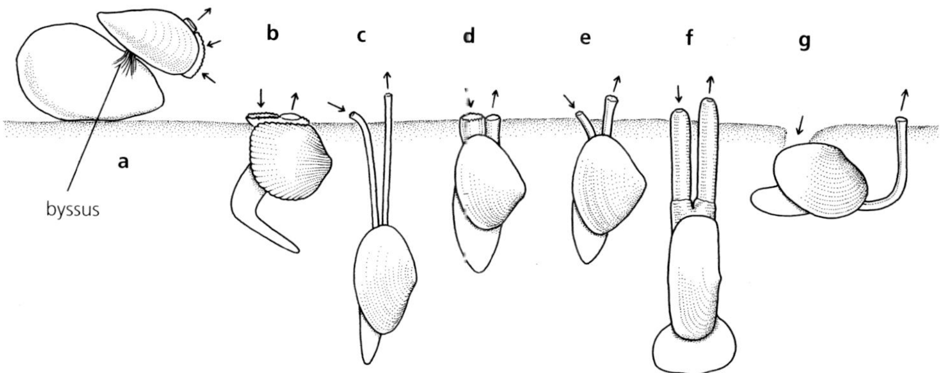

Types of siphons in one surface-living and six burrowing bivalves, showing ingoing and outgoing water currents:
a *sea mussel (Mytilus) — surface-living, attached to a stone by a byssus, with no siphons*
b *heart cockle (Fragum) — shallow burrower with no siphons*
c *tellen (Tellina) — deep burrower with long, separate siphons*
d *venus clam (Katelysia) — shallow burrower with short, joined siphons*
e *mactra (Mactra) — shallow burrower with short, separate siphons*
f *solecurtus clam (Solecurtus) — deep burrower with long, fleshy separate siphons*
g *basket clam (Fimbria) — shallow burrower without an anterior siphon but with a long posterior (outgoing) siphon.*

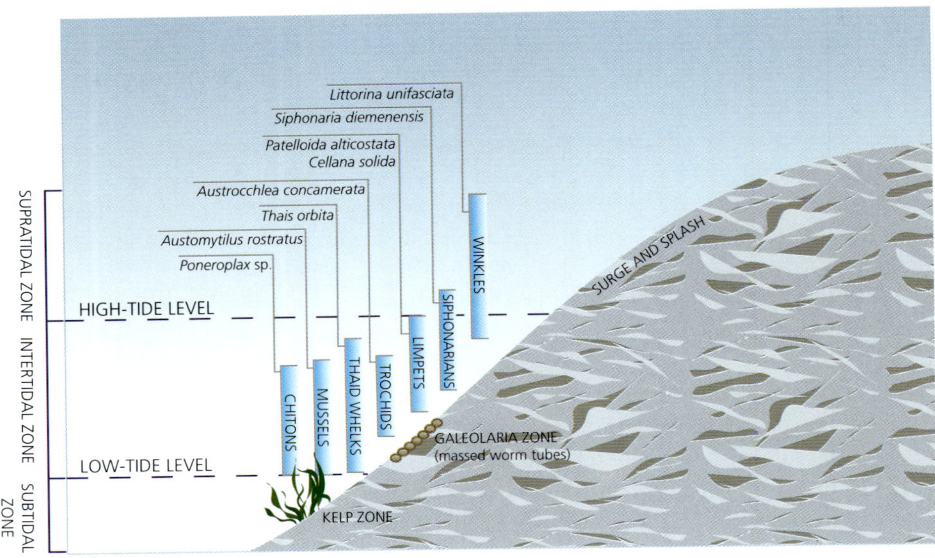

Profile of a typical sloping granite shore in south-eastern Australia showing the different levels (vertical zonation) at which various species of molluscs live.

'LIVING' HABITAT

In many coastal habitats other organisms, either plants or animals, provide habitats for molluscs. Rocky surfaces near the shore are normally vegetated by a variety of algae. These provide diverse habitats for molluscs, including shelter and food for the grazing species, just as grasslands, shrublands and forests do for terrestrial animals. On sandy or muddy seabeds there may be vast meadows of seagrasses providing another range of habitats. A feature of the marine environment that is very different from the land is the common presence of 'forests' of attached invertebrate animals, such as corals and sponges, which also provide shelter and habitat for less sedentary molluscs.

In the tropics, coral reefs provide a variety of structural and biological habitats that support a diverse community of associated animals and plants, including molluscs, known as the coral reef community. These depend for shelter on the physical habitats created by the diverse coral forms, as well as, in the case of coral-eating species, coral tissue for food. Coral reef molluscs include a host of species of colourful shells that are not found in habitats of the more turbid waters of nearby inshore coastal zones. The Great Barrier Reef of the north-east Australian coast needs no introduction and there are extensive coral reefs also on the north-west coast.

Mangroves are trees that grow in the salty water of the upper intertidal zone in the subtropics and tropics. There are several species of mangrove and they form shoreline forests. The largest mangrove forests occur in the mouths of estuaries and in sheltered bays across

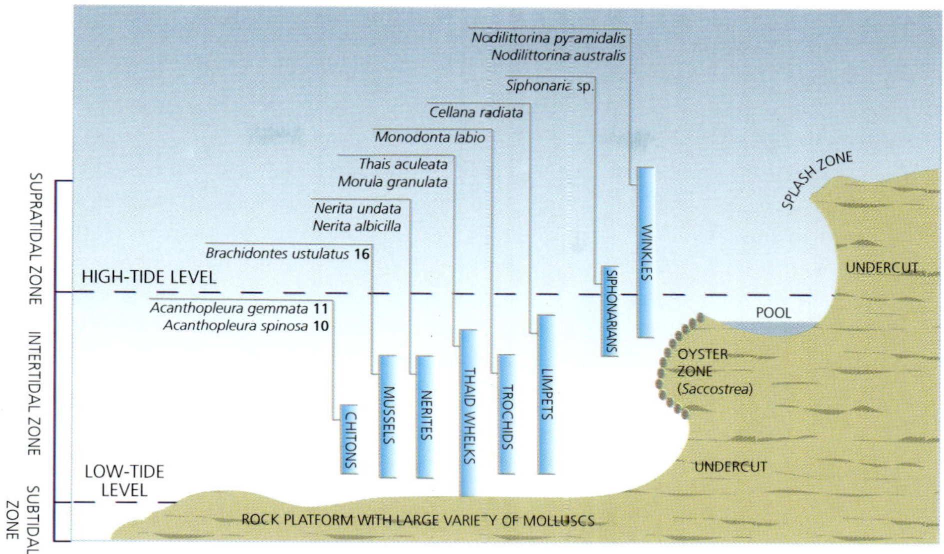

Profile of a typically eroded limestone shore in north-western Australia showing the different levels (vertical zonation) at which various species of molluscs live.

tropical northern Australia. At subtropical latitudes on the east and west coasts there are fewer species of mangrove trees and the invertebrate communities are less species-rich. A characteristic suite of mollusc species (and other invertebrate animals) associate with mangroves. Among the molluscs commonly associated with mangroves are the mud-creepers (*Terebralia*, *Telescopium* and *Cerithidea*), the mangrove littorinids (*Littoraria*) and the Mangrove Nerite (*Nerita balteata*), most of which are not found in other habitats.

SOUTHERN ESTUARIES

On the northern coasts of Australia estuaries are dominated by mud flats and mangrove forests. Southern estuaries do not have mangroves, but they do have their own characteristic faunas, including a few molluscs. The most important physical feature of estuarine habitats is their variable salinity. Generally, estuaries become increasingly less saline further upstream, that is, away from the entrance to the sea. They also vary seasonally, with low salinity in rainy seasons when the rivers that enter them are flowing most strongly. Strongly tidal estuaries may be subject to drastic daily changes in salinity with the ebb and flow of the tide. Most of the invertebrates and fishes that live in the southern estuaries are 'temporary invaders' that enter the estuaries when conditions are favourable but some species are not found elsewhere. Mollusc species that are not found in other habitats include the bivalves *Xenostrobus securis* and *X. inconstans*, and gastropod species of *Salinator* and *Zeacumantus*.

PARASITISM AND COMMENSALISM

Many molluscs depend on other invertebrates for both habitat and sustenance. Egg cowries live among the fronds of soft corals and eat their polyps. Commonly each species of egg cowry has its preferred host soft coral and the relationship is said to be host specific. This is really a specialised form of predation.

Certain species of mussels burrow into colonies of stony corals and use that position for protection and as a convenient site for the capture of planktonic food (perhaps stealing some that the coral might otherwise have caught). This kind of relationship, where one party (the mussel) gains an advantage from the other (the coral) but does it no real harm, is known as commensalism. There are even some parasitic molluscs. One gastropod family, the Eulemidae (not included in this book) comprises small snails that attach themselves to sea stars, sea cucumbers or sea urchins and, through a long snout, suck nutritious juices from the body of the host, as fleas do on a dog.

Feeding

Molluscs consume a vast array of food types and their methods for finding and ingesting them are almost as varied as their shell structures.

Molluscs that feed on fine organic particles are known as detrital feeders. Simple gastropods feed this way by means of a brush-like structure in the mouth, called the radula, that sweeps the particles into the gut. The radula of chitons and most gastropods is modified as a scraping or biting device. It is like a flat, flexible file with rows of tiny chitinous 'teeth'. Special muscles scrape it over the surface of the food, rasping off flakes of tissue which is then bound in mucus and swallowed. As the teeth wear out at the front new ones grow at the back.

Many gastropods are herbivorous and use the radula, modified to a file-like tool, for grazing or browsing on marine vegetation or on the tissues of sessile animals like coelenterate polyps. Some highly specialised predatory gastropods (e.g. cones and augers) and cephalopods (e.g. squids, cuttlefishes and octopuses) have evolved a radula that is a biting or stabbing device. These creatures are efficient hunters with sophisticated sensory and locomotory abilities, and their feeding hardly resembles the detrital gathering of less specialised molluscs.

The bivalves, which have no radula, evolved in another direction altogether. Early bivalves were detrital feeders, sweeping up fine particles by means of flaps associated with the mouth. Later forms caught organic particles in the mucus on their gills. The gills of most modern bivalves are very large, flat ciliated structures arranged like pages in a book and located in a spacious mantle cavity. They function as a net-like filter, catching bacteria or plankton from the water.

Certain bivalve and gastropod families specialise in a parasitic lifestyle. Most parasitic molluscs live on the outer surface of their hosts' bodies, but some live in the gut. (See Parasitism and Commensalism above).

It is hard to imagine that 'dead' shells gathered on the beach once housed living animals with such a wide spectrum of feeding habits.

Reproduction and Development

Have you ever sat on the beach with a shell in your hand and asked yourself where did it come from, how was it created, or why did it die? Just like us, every shell had a mother, and almost every one had a father. The cycle of reproduction is universal but the reproductive and larval development strategies of molluscs are enormously varied. Molluscs reproduce sexually and the sexes are usually separate, individuals being either male or female. However, hermaphroditism, with individuals having both male and female organs, is common.

The life cycle of molluscs follows the normal pattern for animals. In the simplest forms, the females and males mate simply by shedding their eggs and sperm (spawning) into the water, where fertilisation takes place. This is a rather haphazard process. Success depends on there being enough individuals of the two sexes in close proximity and spawning simultaneously, to ensure that the majority of the eggs are fertilised. After fertilisation the egg divides and grows, eventually becoming a larva with all the vital organs and capable of looking after itself. The first stage of larval development in molluscs is called a trochophore. It is followed by a more complex larval form called a veliger that has a minute shell and a ciliated swimming organ or velum. At the critical time, the veliger metamorphoses, with certain drastic changes in body structure and reorganisation of the body organs, and becomes a juvenile, which then grows to adulthood. Enough larvae must survive the terrors of planktonic life to repopulate the adult communities on the seabed. Adults of molluscs that have this style of larval development tend to live in large colonies, at least during the spawning season, and produce large numbers of small eggs.

An alternative strategy is for the eggs to be fertilised within the mother's reproductive tract, and the larvae released after they have already passed through the early development stages, thereby improving their chances of survival. This is called internal fertilisation and is characteristic of nearly all the gastropods (excluding the simplest forms) and the cephalopods.

Internal fertilisation requires one-to-one mating of male and female parents. It also involves behavioural adaptations to ensure that male and female adults meet, recognise and accept each other at appropriate times. It may come as a surprise, but the creature that made that small gastropod shell in your hand once had a love life.

In the chitons and the simplest bivalves and marine gastropods, planktonic larval development usually takes a few days, and the larvae depend on nourishment supplied in the egg by the mother. These larvae are unable to disperse widely with the ocean currents. In more specialised marine molluscs the veliger larvae become capable of feeding themselves in the plankton, usually by capturing minute particles such as bacteria. Species that do this may be able to sustain the planktonic phase for long periods. Gastropods that have long-lived self-feeding larvae include tritons, trumpets and helmet shells. Their veligers may survive for many months in the plankton, often distributed across wide areas of the oceans and capable of colonising suitable habitats far from the site occupied by their parents.

Many gastropods and cephalopods protect the early development stages within capsules until their development is advanced. A jelly-like substance is folded around batches of eggs just before they emerge from the mother's reproductive opening and solidifies into stiff capsules when it meets the sea water. The capsules containing the eggs are then laid in an egg mass, usually attached to firm objects on the seabed. Commonly, the larvae hatch from the egg capsules as swimming, late-stage veligers immediately capable of feeding themselves. This strategy makes the best of both worlds — the early larval stages are protected, yet the later stages can take advantage of the rich food reserves of the ocean and the potential for wide dispersal in the water currents.

The veliger larvae of some gastropods, such as volutes, complete their development within the egg capsules and hatch as miniature snails, without having any planktonic dispersal phase. This is known as direct development. Species which adopt direct development maximise the protection side of the equation and forfeit the advantages of dispersal.

Yet another reproductive variation is for the female to brood the larvae in the mantle cavity. Many bivalves and gastropods do this, achieving the same result as laying the fertilised eggs in capsules, that is, providing the young molluscs with a good start to life by protecting them during their early development.

Collecting Shells

Making a shell collection can be a very satisfying activity. It can lead to an interest in natural history or even to serious study. However, collectors need to be conscious of the local impact of their activities and take great care not to damage habitat.

CONSERVATION

There are no known cases of marine molluscs becoming extinct in modern times and few, if any, strictly marine species are in danger of extinction. However, while populations in deeper water may not be threatened, some species may be locally vulnerable. For example, local populations of volute species that inhabit sandy cays in Queensland, and the *Zoila* cowries of Western and South Australia, have disappeared, possibly due to collecting. The conservation issues here are about disruption to local ecosystems and the loss of particular genetic strains, rather than the threat of species extinction.

Another conservation issue is chronic pollution of inshore bays and estuaries that may affect whole communities of plants and animals. This is a major issue in southern Australian estuaries, where some estuarine species, including some molluscs, are at risk of extinction if their sole habitat becomes uninhabitable.

Collectors should be aware that there are regulations that prohibit or restrict the collecting of live specimens in many parts of Australia. Some molluscs, such as the giant clams (*Tridacna*) are totally protected. There are special conservation areas around the coast where all plants and animals are protected, or where restrictions apply and you may require a collecting permit. Generally speaking, collect only empty ('dead') specimens so as to avoid having an ecological impact. Leave any shells that have been re-occupied by hermit crabs or other creatures. If you must take live specimens, find out from conservation and fisheries authorities where this is allowed and what formalities may apply.

People sometimes covet live-collected specimens because their shells are in good condition, but you must remove the animal before storage and that can be difficult. Bodies of dead molluscs have a particularly offensive smell! Far too often, collectors full of good intentions take live specimens only to discard them when they become smelly. You should take live specimens only if it is legal, and only if you really need them for study purposes and you have the facilities to properly clean them. Otherwise small lives may be wasted.

PREPARING SHELLS FOR THE COLLECTION

Empty shells are usually easy to prepare for the collection — simply wash them in fresh water. Treatment with a light oil (vaseline or 'baby oil') will enhance their colours.

Live-collected bivalves are relatively easy to 'clean' as they usually pop open when put into hot water for a few minutes, after which you can scrape the body out. Gastropods are more difficult. If it is essential to take live gastropods for specimens the best method of removing the dead body is to keep it damp (not wet) and allow it to decompose to the point where you can shake or wash out the body with a water jet from a hose. Burying the shell in clean, damp beach sand for a few days or repeated freezing and thawing can also do the trick.

Serious collectors always keep locality and habitat data with their specimens. You can tuck labels with names and collecting data into the aperture of gastropods. Keep small specimens in separate small plastic bags or vials, each with their own label.

Dangerous Molluscs

The notion that a spineless snail could be dangerous may come as a surprise. However, certain of the predatory gastropods are equipped with poison glands and means of delivering the poison, which is used to subdue prey. The most infamous of these are the cone shells (family Conidae). Collectors are strongly advised to leave live cone shells alone.

All cones have a poison gland derived from the salivary gland in the mouth. The radula also is highly modified in this family and comprises a bundle of spear-like shafts that are held in a sac where they are bathed in the poison. To spear prey, the animal thrusts the shafts out through its mouth by means of vigorous muscular contractions. Once the prey has succumbed to the poison, it is swallowed whole.

Most cones prey on marine creatures such as worms and other molluscs and their sting has only a mild effect on humans, about equivalent to a bee sting. However, a few fish-eating cones have an ultra-powerful venom and these are very dangerous. Several fish-eating species have been responsible for human fatalities, including some in Australia. The two cones known to have caused fatal attacks are *Conus geographus* and *C. textile*, but others suspected of being dangerous are *C. striatus*, *C. tulipa*, *C. catus*, *C. omaria*, *C. imperialis* and *C. lividus*.

Another group of dangerous marine molluscs common around Australian shores are the blue-ringed octopuses (genus *Hapalochaena* — see picture page 177). In this case the poison is administered by means of a bite from beak-like jaws. Not all octopuses have a bite that is dangerous to humans but the blue-rings are especially dangerous and have been responsible for several Australian fatalities. While octopuses are not shelled molluscs (and perhaps beyond the scope of this book) they are often met by collectors because of their habit of using the empty shells of other molluscs as temporary homes. People picking up supposedly empty shells in tide pools need to take great care, as they may have collected more than they intended.

This is not the place for advising on treatment of cone and blue-ringed octopus stings. If stung seek professional medical attention immediately. The best policy is to avoid the risk by not handling live specimens.

Live Striate Cone (Conus striatus) *with egg mass.*

The Shells

Chitons   Class POLYPLACOPHORA

Chitons, also known as coat-of-mail shells, have a flat-soled body protected by a shell comprising eight articulating parts, called plates, embedded in a tough, scaly or spiny surrounding 'girdle'. Their anatomy is quite simple compared to that of gastropods. Most are very slow moving and live on hard surfaces such as rocks. There are nine families on Australian shores. A few common species are illustrated here to represent the class.

TORR'S ISCHNOCHITON *Ischnochiton torri*
Ischnochitons are mostly intertidal and shallow subtidal animals characterised by a scaly girdle and finely sculptured valves. There are many species in southern Australia. This one grows to 40 mm long. Rounded, orange-brown back with a cream axial stripe; rather narrow, minutely scaled brown girdle. Very common under stones. Widespread across southern Australia.

CORRODED ISCHNOCHITON *Ischnochiton cariosus*
Pale and straw-coloured. Grows to about 35 mm long. Very common under stones, often with Torr's Ischnochiton. Southern Australia and up the west coast as far north as Shark Bay.

LINED ISCHNOCHITON *Ischnochiton lineolatus*
Elongate species with a white back bearing axial brown dashes. Grows to 40 mm long. Common under stones in the intertidal zone from Bass Strait to the central west coast.

PUSTULOSE ISCHNOCHITON *Ischnochiton contractus*
Similar to the Lined Ischnochiton and about the same size but broader and with a row of wide dark brown blotches along the back. Common under stones in the intertidal zone from Bass Strait to the central west coast.

VERCO'S ISCHNOCHITON *Ischnochiton verconis*
Light brown. Broad for the genus — an adult 50 mm long would be about 40 mm wide. Under stones in the intertidal zone. South-western Australia.

GREENISH ISCHNOCHITON *Ischnochiton subviridis*
Bluish green valves with white blotches on the back. Dark brown blotches around the narrow girdle. Smaller than most in the genus; grows to about 30 mm long. Very common under stones in the intertidal zone. Relatively restricted geographic distribution, only in Bass Strait.

STRIATE LEATHERY CHITON *Cryptoplax striata*
One of a group of elongate, almost worm-like chitons. Their tiny shell valves do not articulate with each other, but are set in a thick, leathery brown girdle. Grows to 120 mm long. Under stones and squeezed into crevices in the intertidal and subtidal zones. Southern Australia.

Corroded Ischnochiton (left) and Torr's Ischnochiton (right)

Lined Ischnochiton

Pustulose Ischnochiton

Verco's Ischnochiton

Greenish Ischnochiton and Striate Leathery Chiton

PALE-FLAMED CHITON *Plaxiphora albida*

Rounded, varicoloured back. Wide girdle bearing masses of short needle-like structures. Grows to about 70 mm long. Attached to rocks in the lower part of the intertidal zone. Southern Australia.

MUD CHITON *Lorica cimoliak*

Light brown with keeled shell valves along the back. Large; grows to 90 mm long. Mainly in the subtidal zone; seems to prefer muddy water. Southern Australia.

SPINY CHITON *Acanthopleura spinosa*

A large black species easily recognised by the long spines on the girdle. Grows to 100 mm long. Very conspicuous on rocks in the upper intertidal zone of the tropics. Northern Australia.

GEMMATE CHITON *Acanthopleura gemmata*

Resembles the Spiny Chiton, about the same size and found in the same places but differs in that the girdle spines are numerous but minute. The back is usually brown. Common on rocks in the upper intertidal zone. Northern Australia.

BLACK AND WHITE CHITON *Clavarizona hirtosa*

This shell has thick, greenish plates with dark brown margins and an axial brown band. The back is often eroded so that the original colour is obscured. The narrow, pustulose girdle has alternating grey and black patches. Grows to about 60 mm long. Common, usually gregarious, in a distinct band just below the limpet zone in the middle intertidal zone of rocky shores. South-western Australia.

Pale-flamed Chiton

Mud Chiton

Spiny Chiton

Gemmate Chiton

Black and White Chiton

Bivalves Class BIVALVIA

There are five subclasses in this class of molluscs and about 80 families represented in the Australian marine fauna. Some are deep-sea creatures not seen in shallow coastal waters. The 23 families representing the class in this book have been selected because they include species whose shells are commonly found on Australian beaches, which live in easily accessible intertidal or shallow subtidal habitats or which are of special interest for some reason. Together they cover the diversity of form and lifestyles that characterise this large group.

SEA MUSSELS Family MYTILIDAE

Sea mussels may be attached by hairy byssal threads to hard substrates, burrow in sand or mud or bore in calcareous rocks. Attached forms may live gregariously in vast 'mussel beds' on the sea floor or in the intertidal zone. Throughout the world large mussels are fished commercially for food. There are about 70 species in Australian waters. The shells are usually long and inequilateral: that is, most of the growth is posterior so that the umbos are at or near the narrow front end. Hinge teeth are small or lacking but there may be a series of small teeth along the margin behind the long ligament, which are known as pseudotaxodont teeth.

SOUTHERN HORSE MUSSEL *Modiolus areolatus*
Mussels of this genus have no hinge teeth, smooth, unsculptured shells and a thick, hairy periostracum. This common species is brown and has a narrow, pouted anterior end. Grows to 100 mm long. Attached singly or in clusters. New Zealand and southern Australia.

PHILIPPINE HORSE MUSSEL *Modiolus philippinarum*
Like the Southern Horse Mussel but wider at the front and much larger. Named after the place from which the original specimens came. One of the largest of the sea mussels; grows to more than 120 mm long. Very common on intertidal flats where it attaches to shells and small stones in the sand. Common throughout much of the Indo-Pacific and northern Australia.

EARED HORSE MUSSEL *Modiolus auriculatus*
Smaller than *M. philippinarum* — grows only to about 50 mm long — but of similar form except for its pronounced compressed 'ears' at the posterodorsal corner. Rocky or coral flats in the intertidal zone. Widespread throughout the Indo-Pacific; northern Australia.

WINGED HORSE MUSSEL *Modiolus micropteris*
Nearly triangular outline with a pronounced dorsal angle forming a 'wing'. Smooth, slightly shiny light brown exterior; white interior but purple in the dorsal part. Grows to about 10 cm long. Common on mud flats. Indo-Pacific and northern Australia.

Southern Horse Mussel

Philippine Horse Mussel

Eared Horse Mussel

Winged Horse Mussel

ROUGH BEAKED MUSSEL *Brachidontes erosa*

Brachidontes shells have a strong sculpture of radial ribs. The hinge is toothed anteriorly and there are small pseudotaxodont teeth behind the ligament. In this species the shells are almost cylindrical. Bluish grey underneath, purple on the upper half with a thin brown periostracum. Grows to 80 mm long. Attaches to stones on intertidal sand and mud flats, estuaries and sheltered bays. Common species of southern Australia.

MYSTERY MUSSEL *Brachidontes ustulatus*

So named because its identity was confused for many years. More compressed and finely ribbed and a little smaller than *B. erosa*. Both species occur on the south coast of Western Australia but in different habitats. *Brachidontes ustulatus* has a bluish grey exterior, brown on the umbo and on the posterior end. Grows to about 60 mm long. Lives on rocky shores in more open conditions where it sometimes covers large areas in colonies which may be peeled off the rock like a mat. It has an unusual distribution from South Australia to the Kimberley.

BLUE SEA MUSSEL *Mytilus edulis planulatus*

Common mussel fished and cultivated commercially in southern Australia. Closely related mussels occur in similar latitudes elsewhere in both the Northern and Southern Hemispheres — hence the use of the subspecies name. Blue-black with a pointed anterior end, a smooth exterior and small anterior hinge teeth. Grows to 120 mm long. Attached to rocks or jetty pilings in the lower intertidal and shallow subtidal zones of sheltered bays and estuaries. Southern Australia.

LITTLE BROWN MUSSEL *Xenostrobus securis*

Elongate and compressed with a smooth brown shell which has no hinge teeth. Grows to about 40 mm long. One of the few obligate estuarine molluscs in southern Australia. Lives in colonies in the upper parts of estuaries where the water may be virtually fresh for part of the year. Southern Australia including the eastern and western overlap zones. Also in New Zealand.

VARIABLE BROWN MUSSEL *Xenostrobus inconstans*

Easily confused with *X. securis* but more slender and subcylindrical. Shiny brown or bluish. Grows to 30 mm long. Intertidal flats in the middle parts of estuaries from Tasmania to Albany.

BLACK DATE MUSSEL *Lithophaga teres*

Date mussels are so named for their elongate–cylindrical shells that are said to resemble the seeds of dates. They are borers. This common species is sculptured anteroventrally by transverse cords and has a toothless hinge. Smooth, shiny black-brown periostracum beneath which the shell is white. Grows to 75 mm long. Bores in dead corals and limestone rocks. Indo-Pacific and northern Australia extending far south into the temperate zone on both sides of the continent.

Rough Beaked Mussel

Mystery Mussel

Blue Sea Mussel

Little Brown Mussel

Variable Brown Mussel

Black Date Mussel

BROWN DATE MUSSEL *Lithophaga nasuta*

One of several Australian species of this genus with brown shells and no sculpturing. Instead there is a superficial layer of calcareous encrustation (accretion) over the outer surface. Grows to over 60 mm long. Bores into limestone rocks in the intertidal zone. Common across the northern coast.

POINTED DATE MUSSEL *Lithophaga mucronata*

Slender and thin-shelled with a pointed posterior end bearing a thick, chalky projecting accretion. Light brown. Grows to about 25 mm long. Common species that bores in limestone and dead coral in the intertidal and shallow subtidal zones. Indo-Pacific and northern Australia.

HAIRY MUSSEL *Trichomya hirsutus*

Solid brown mussel with beaked anterior umbos and a thick outer covering of serrate periostracum. Grows to about 60 mm long. Attaches by a strong byssus to rocks in the intertidal and shallow subtidal zone. Eastern Australia.

SENHOUSE'S MUSSEL *Musculista senhousia*

A delicate little mussel that has invaded Australian bays and estuaries in recent years. Believed to originate in South-East Asia. While alive the greenish or light brown weakly sculptured shell is enclosed in a bag-like nest of byssal threads and mucus. Grows to about 30 mm long. Muddy sand in the shallow subtidal zone. Now established in southern and eastern Australia.

ARK SHELLS Family ARCIDAE

The shells of this large family are usually equivalve and inequilateral with the umbos anterior to the mid-line. Taxodont teeth, arranged in an arched series along the hinge, may be straight or chevron-shaped and are often smaller at the middle. There is a dense brown fibrous or hairy periostracum. Most attach by a strong byssus to hard objects but some lie free in mud or sand.

INDO-PACIFIC ARK *Arca navicularis*

Rectangular with a straight dorsal margin and prominent lateral wings. Large, wide apart umbos. Narrow ventral gape for the byssus. About 120 small, subequal teeth on the hinge. Coarsely ribbed cream exterior with a pattern of reddish brown wavy lines. Grows to 80 mm long. Attaches to underside of rocks and coral slabs. Indo-Pacific and north-eastern Australia.

VENTRICOSE ARK *Arca ventricosa*

Elongate, winged shell with large, wide apart umbos. Unlike the species above, the ribs are fine. Ventral side is convex with a very wide lens-shaped byssal gape. Straight hinge with about 90 nearly vertical teeth. Exterior off-white to light brown anteriorly, brown posteriorly; brown ribs. Brown transverse lines across the dorsal area are a distinguishing feature. Grows to 100 mm long. Tightly adheres in crevices of massive corals such as *Porites*. Indo-Pacific and northern Australia.

Brown Date Mussel

Pointed Date Mussel

Hairy Mussel

Senhouse's Mussel

Indo-Pacific Ark

Ventricose Ark

HAZELNUT ARK *Arca avellana*
Similar to other species of the genus (previous page) but with a smaller byssal gape and the dorsal area between the umbos is uniform brown. Grows to 40 mm long. Attaches to rocks and crevices. Indo-Pacific and northern Australia.

TWISTED ARK *Trisidos tortuosa*
The incredibly twisted shell makes this one of the most remarkable bivalves in the region. The creamy, brown-flecked, finely corded valves are thin, laterally compressed and elongate. Straight hinge with 9–12 large, oblique teeth at the ends and numerous tiny teeth centrally. Grows to 100 mm long. Also known as the Propella Ark. Burrows just below the surface in muddy substrates in the shallow subtidal zone. Northern Australia.

GRANULAR MUD ARK *Anadara granosa*
The mud arks are remarkable for their red blood. There are several species in northern Australia and one in the south. They have thick, white, inflated shells bearing a thick, furry periostracum. This species has a thick, almost equilateral shell with strong, granose ribs. Grows to 60 mm long. Mud, often associated with mangroves, in the intertidal zone. In its preferred habitats it may be extremely abundant and is commercially harvested for food in Asia. Indo-Pacific and northern Australia.

TRAPEZOID MUD ARK *Anadara trapezia*
Larger than the Granular Mud Ark with less-granose ribs. White beneath a brown, furry periostracum. Grows to 70 mm long. Sand and muddy areas of bays and estuaries along the east coast. In the Late Pleistocene this species was also common along the west coast but is now extinct in Western Australia except for a remnant population in Oyster Harbour at Albany.

FINELY RIBBED MUD ARK *Anadara crebricostata*
Relatively thin-shelled and more finely ribbed than most of the other Australian species. Rather elongate. White beneath a brown, furry periostracum. Grows to about 75 mm long. Sandy flats across the northern Australian coast.

SOUTHERN BEARDED ARK *Barbatia pistachia*
In this genus the exterior bears a strong, hairy periostracum. This very common species has a more or less oblong, laterally compressed shell with rounded ends and a fine sculpture of radial and concentric cords. Usually off-white with pale brown bands. Grows to about 60 mm long. Attaches in crevices and on the undersides of stones. Southern Australia.

ALMOND ARK *Barbatia amygdalumtostum*
Rectangular with rounded ends. Dark purple-brown both inside and out. Grows to 70 mm long. Attaches under stones. This is perhaps the most common species of the genus in northern Australia.

Hazelnut Ark

Granular Mud Ark

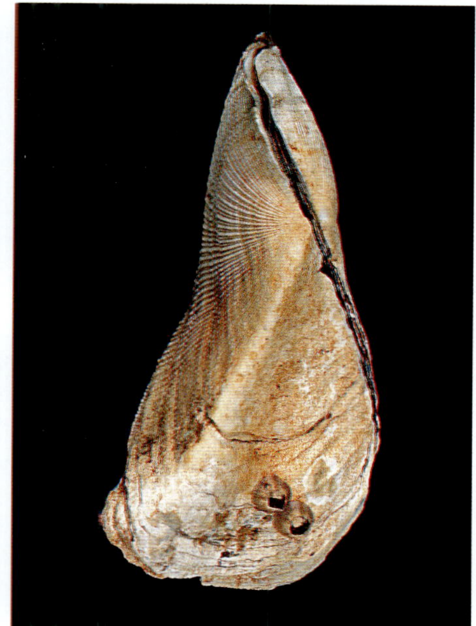

Twisted Ark

Trapezoid Mud Ark

Finely Ribbed Mud Ark

Southern Bearded Ark

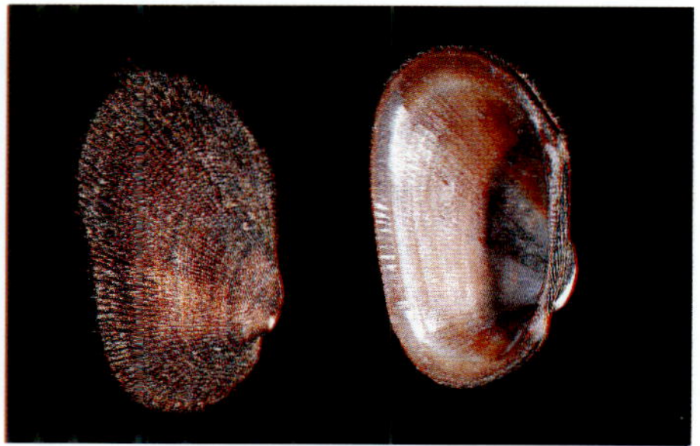

Almond Ark

BITTERSWEET CLAMS Family GLYCYMERIDIDAE

Most of these bivalves have thick, compressed shells that are more or less circular and equilateral. They have an arched hinge bearing taxodont teeth and in that respect they resemble the ark shells. There are three genera in Australia with more than 20 species.

GRAY'S BITTERSWEET CLAM *Glycymeris grayana*
The shells of this genus are finely ribbed. This species has a light brown, almost smooth exterior with white blotches and a velvety periostracum. Grows to 50 mm diameter. Shallow subtidal zone. Often cast ashore on oceanic beaches of south-eastern Australia.

STRIATE BITTERSWEET CLAM *Glycymeris striatularis*
Distinguished from the previous species by its more uniform brown colour and stronger ribs comprising bundles of 6–10 ultra-fine cords. Periostracum hairy rather than velvety. Grows to 45 mm diameter. Another common southern Australian species also often found on oceanic beaches.

COMB BITTERSWEET CLAM *Tucetona pectunculus*
Much heavier ribs than *Glycymeris* and with a distinctive pattern of toothed, concentric brown lines. Grows to 45 mm diameter. Common on coral reefs of the Indo-Pacific and Northern Australia where it lives in sand on reef flats in the shallow subtidal zone.

RAZOR CLAMS Family PINNIDAE

Large, equivalve and very inequilateral shells with terminal umbos and pointed anterior ends. The long, straight hinge is toothless and the ligament is long and thin. A portion of the shiny interior surface is mother-of-pearl. In Australia there are three genera and eight species. They live rooted in sand or mud attached by a large byssus to stones or shell fragments, with the pointed anterior end downwards and the wide, sharp-edged posterior end projecting above the surface of the sediment. They are called razor clams because of the shells' sharp-edged posterior margins. Beware of treading on them in bare feet in the seagrass at low tide! Some species are fished for food in parts of the Indo-Pacific but not commercially. In previous times the byssal threads of the Mediterranean *Pinna nobilis* were gathered and spun into cloth.

COMMON RAZOR CLAM *Pinna bicolor*
In *Pinna* the upper and lower lobes of the internal mother-of-pearl area inside each valve are separated, with the adductor muscle scar located entirely within the upper lobe. In this species the posterior margin is convex, the exterior brown with pale radiating rays and sculptured with spiny radial ribs. Grows to 500 mm long. Very common on tidal flats. Often associated with seagrass beds in southern Australia. Widely distributed in the Indo-Pacific. Circum-Australian.

Gray's Bittersweet Clam

Striate Bittersweet Clam

Comb Bittersweet Clam

Common Razor Clam

DELTOID RAZOR CLAM *Pinna deltodes*

Some say this is a form of the Common Razor Clam. It differs in being fan-shaped, more compressed and stout and sculpture is lacking or weak. Grows to about 400 mm long. Intertidal sandy flats. Northern Australia.

PRICKLY RAZOR CLAM *Pinna muricata*

Distinguished by its more triangular outline with a blunt posterior margin and spiny radial ribs. Horn coloured with brown or sometimes bluish brown blotches. Grows to about 300 mm long. Sandy tidal flats. Widespread in the Indo-Pacific; in Australia it is confined to the northern coast.

COMB-LIKE RAZOR CLAM *Atrina pectinata*

Atrina is distinguished from *Pinna* by an undivided mother-of-pearl layer. This species is horn-coloured and thin-shelled with imbricate ribs, blunt dorsal margin and triangular outline. It gets its common name from the row of tooth-like spines that are often present on the main dorsal rib. Sand or muddy sand from the intertidal zone to considerable depths. A species from the central Indo-Pacific that is common also across northern Australia.

TASMANIAN RAZOR CLAM *Atrina tasmanica*

Thin, fragile shell with a rounded posterior margin and strongly imbricate ribs. Horn-coloured. Grows to 250 mm long. Sand in the intertidal and subtidal zones. Southern Australia.

WING OYSTERS AND PEARL OYSTERS Family PTERIIDAE

This diverse group of bivalves ranges from small, thin-shelled *Electroma* species (not illustrated) to the thick-shelled pearl oysters of *Pinctada*. All share a spectacular mother-of-pearl interior; some are famous for their pearls. The left valve is more inflated than the right. There are about eight species of *Pinctada* in Australia, about 15 *Pteria* and at least eight *Electroma*.

GOLD-LIP PEARL OYSTER *Pinctada maxima*

Pinctada valves are almost equivalve and lack a prominent posterior wing. This is the largest species of the genus and the one that is most fished and cultured commercially for its 'mother-of-pearl' and its pearls. Light brown or greenish scaly exterior. Lustrous interior with a golden zone around the perimeter. Grows to nearly 300 mm long. In its natural habitat this oyster lives attached by a strong byssus to a rocky substrate from the intertidal zone down to depths of many metres. Indo-Pacific and northern Australia.

BLACK-LIP PEARL OYSTER *Pinctada margaritifera*

Smaller than the above species; grows to about 170 mm long. Dark brown or greenish exterior with prominent scales. The internal mother-of-pearl is often tinged with rose and its margins are black. This species is also fished and cultured commercially. Attaches to rocks at depths of at least 30 m. Widespread in the Indo-Pacific region and common across northern Australia.

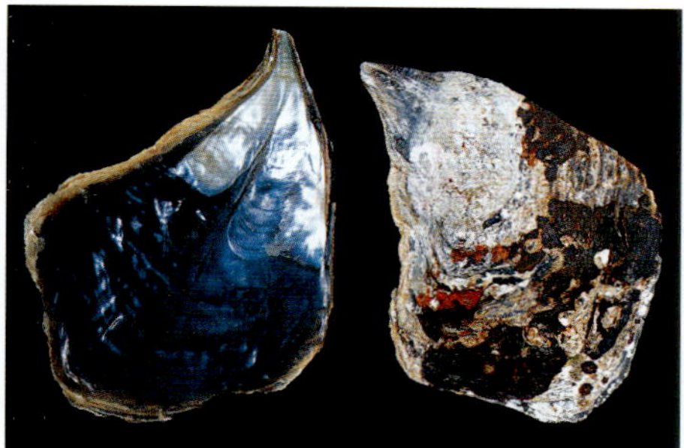

Deltoid Razor Clam

Prickly Razor Clam

Comb-like Razor Clam

Tasmanian Razor Clam

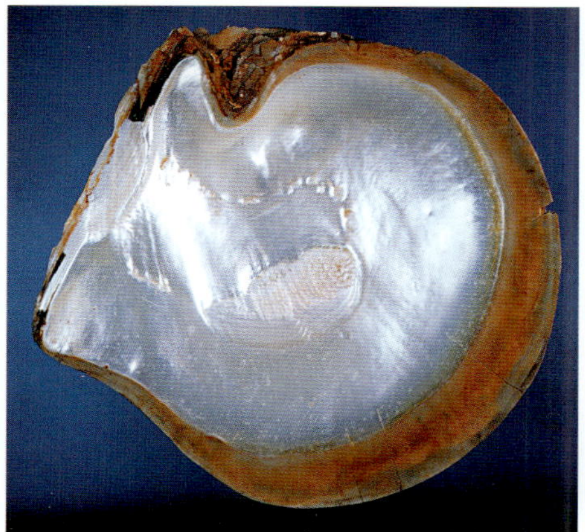

Gold-lip Pearl Oyster

Black-lip Pearl Oyster

PALE PEARL OYSTER *Pinctada albina*

One of the smaller species of the genus growing to about 100 mm long. Yellowish exterior with weak scales. Golden internal mother-of-pearl. Abundant in the seagrass beds of Shark Bay, Western Australia, where it was once fished commercially. Today it is a culture species there. Northern Australia.

PENGUIN WING OYSTER *Pteria penguin*

The valves of this genus are inequivalve and the posterior margin is extended to form a long wing. This large species has a dark exterior and a brilliant silvery internal mother-of-pearl. Grows to at least 200 mm long. Attaches to rocks and sea whips in the shallow subtidal zone. Widespread in the Indo-Pacific and across northern Australia.

RED WING OYSTER *Pteria lata*

Thin shell with a short anterior wing and a pronounced posterior wing. Young individuals like the figured specimen may be elongate. Red exterior with a periostracum of fine, short hairs. Attaches to sea whips in the subtidal zone. Grows to 140 mm long. Widely distributed in the Indo-Pacific and across northern Australia.

TOOTHED PEARL SHELLS, HAMMER OYSTERS AND SPONGE FINGERS Families ISOGNOMONIDAE and MALLEIDAE

The two families in this grouping both have a strong byssus and partly mother-of-pearl interior. Toothed pearl shells (family Isognomonidae) are tropical and warm temperate bivalves. Their straight hinge is toothless but has a series of short ligamental pits that resemble a row of teeth; these give the family its name. Hammer oysters and sponge fingers (family Malleidae) are similar, but the hinge is simple and there is a single, triangular ligamental area. In the genus *Malleus* lateral extensions of the hinge form pronounced wings from which comes its common name. The wingless shells of the genus *Vulsella* are called sponge fingers.

PACIFIC TOOTHED OYSTER *Isognomon isognomon*

The shells have straight or curved, more or less parallel sides; hinge length is about equal to the width of the shell. A short posterior wing is sometimes differentiated. Purple, black or brown scaly exterior. Large internal mother-of-pearl area; narrow non-mother-of-pearl border. Grows to 100 mm long. Clusters or singly under stones and in crevices. Widespread in the Indo-Pacific and across northern Australia.

IRREGULAR TOOTHED OYSTER *Isognomon legumen*

Like its cousin *I. isognomon,* and of similar size, but of very irregular outline and the hinge is shorter than the widest part of the shell. Internal mother-of-pearl area is narrower than the non-mother-of-pearl border. Brown to grey exterior sometimes with darker radial lines. Clusters or singly under stones and in crevices. Common in the Indo-Pacific and northern Australia.

Pale Pearl Oyster

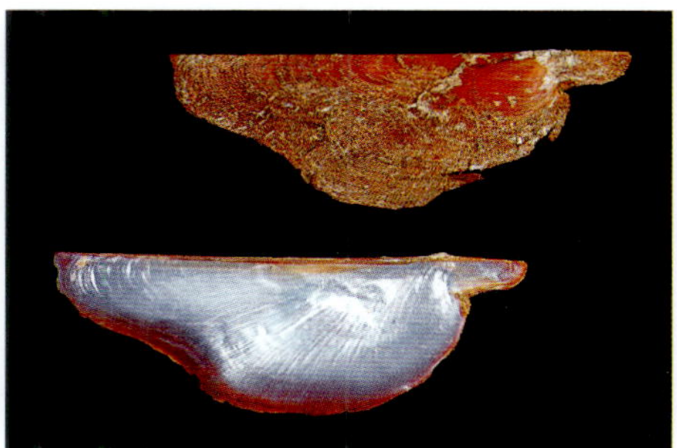

Red Wing Oyster

Penguin Wing Oyster

Pacific Toothed Oyster

Irregular Toothed Oyster

LITTLE TOOTHED OYSTER *Isognomon nucleus*

Small subquadrate shell beaked with a prominent byssal notch. Smooth grey to black exterior is often eroded. Internal mother-of-pearl area is tinged with purple and has a black margin. Grows to 30 mm high. Rock crevices and among rock oyster clusters or mangrove roots in the upper intertidal zone. Indo-Pacific and northern Australia.

WHITE HAMMER OYSTER *Malleus albus*

Has a long 'shaft' with rather straight but fluted sides. Prominent wings; anterior wing is shorter. Yellow to grey-white exterior with pale brown blotches. Bright, silvery internal mother-of-pearl area. Grows to 180 mm high. Attaches to hard objects on sand or gravel substrates. Indo-Pacific and northern Australia. A similar Hammer Oyster, *Malleus meridianus*, lives in southern Australia — it is darker coloured and almost as tall but its wings are of almost equal length.

SPONGE FINGER *Vulsella vulsella*

Thin, compressed shells with a variable outline. Usually tall (up to 80 mm) and lacking wings. Yellow-brown exterior faintly sculptured with scaly radial cords. Interior almost entirely covered by silvery mother-of-pearl. Embedded in sponges; often found in sponges cast up on beaches. Seemingly everywhere throughout the Indo-Pacific. Circum-Australian.

FILE SHELLS Family LIMIDAE

File shells are equivalve and higher than long. There may be small taxodont hinge teeth, but the hinge is usually toothless and the pointed umbos are widely separated. Some species live attached by a byssus but others are free-living and able to swim by means of the rowing action of long mantle tentacles. This small family has a long fossil history and six living genera. There are about 20 Australian species.

SPINY LIMA *Lima vulgaris*

Strong white compressed shell, obliquely inequilateral with a sculpture of spiny radial ribs. Grows to 65 mm high. Attached by a byssus under stones and is common in the intertidal and subtidal zones of rocky shores and coral reefs around Australia. Cosmopolitan.

FRAGILE LIMA *Limaria fragilis*

Fragile as its name indicates, tall and obliquely inequilateral, the shell gapes both in front and behind. Fine sculpture of radial cords. White; the animal has a bright red mantle and is capable of swimming. Grows to 25 mm high. Common under stones and slabs of rocky shores and coral reefs. Widespread in the Indo-Pacific and northern Australia.

White Hammer Oyster

Little Toothed Oyster

Sponge Finger

Spiny Lima

Fragile Lima

OYSTERS Families OSTREIDAE and GRYPHAEIDAE

These two families are distinguished by the shape of the muscle scar, which is flattened or kidney-shaped in the Ostreidae and circular in the Gryphaeidae (the latter represented here only by the genus *Hyotissa*). The interior surface of these oysters is not mother-of-pearl. Often the left valve is more concave than the right valve. The hinge of adults is toothless and there is a single adductor muscle. Some species live free on the surface of the sediment but most cement themselves to a firm substrate (by the left valve). In northern Australia there are several species of rock oyster that live in aggregated masses forming distinct intertidal 'oyster zones' along rocky shores. The gender of some oysters alternates from male to female.

SOUTHERN FLAT OYSTER *Ostrea angasi*
Also known as the Mud Oyster. Commercially harvested in southern Australia and was an important food of Aborigines in pre-European times. Rough, scaly surface; fluted margins. Off-white to horn-coloured. Grows to 120 mm high. Free-living on muddy flats. Southern Australia.

SYDNEY ROCK OYSTER *Saccostrea glomerata*
White shells with an irregular, trapezoidal outline and very unequal valves. The upper valve is flattened, fitting within the convex left valve which is cemented to a hard substrate in the shallow subtidal zone. Cultivated commercially. Grows to about 100 mm high. Eastern Australia.

HOODED ROCK OYSTER *Saccostrea cucculata*
Like the Sydney Rock Oyster but has a very irregular shape due, in part, to the way these oysters crowd together in thick bands in the intertidal zone. White interior with a purple-black margin. Sometimes harvested commercially. An Indo-Pacific species abundant across northern Australia.

COCK'S-COMB OYSTER *Lopha cristagalli*
A large (up to 120 mm) oyster, subcircular in outline, with four to eight very strong, acute, thick folds that interlock at the deeply serrate margins. Lilac or brown exterior; bronze interior. However, the shell is usually covered by sponges. Lightly attaches by clasper spines to stones or other hard objects. Another Indo-Pacific oyster common in northern Australia.

FRONS OYSTER *Dendrostrea frons*
Elongate–oval, both valves convex and strongly folded; upper valve sometimes has a longitudinal ridge. White to brown, the shiny inner surface taking the hue of the exterior. Grows to about 80 mm long. Attaches to rocks or gorgonians. Indo-Pacific and northern Australia.

HONEYCOMB OYSTER *Hyotissa hyotis*
Huge, heavy shell with strong, spiny, radial folds and an irregular margin. There are fine transverse ridges below the ligament. Grey or olive-brown exterior; white interior with black edges. Grows to 20 cm long. Lightly cemented to stones in the shallow subtidal zone. Indo-Pacific and northern Australia.

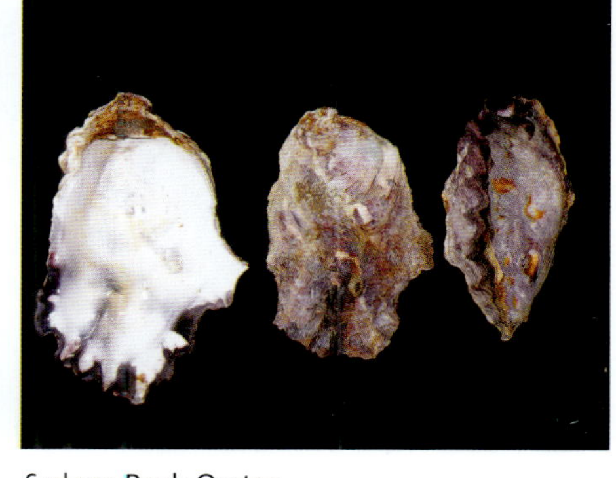

Southern Flat oyster

Sydney Rock Oyster

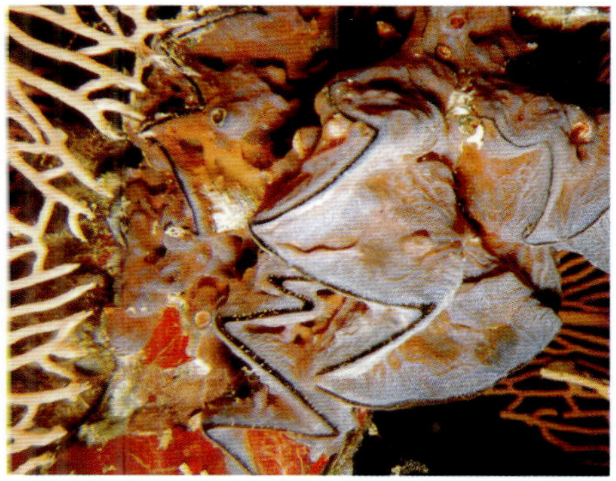

Hooded Rock Oyster

Cock's-comb Oyster (in life attached to a gorgonian)

Frons Oyster

Honeycomb Oyster

SCALLOPS Family PECTINIDAE

Scallops are among the best known of the bivalves because of their colourful shells and because some are good to eat. There is a single large, central muscle scar. The interior is not mother-of-pearl. Most have inequivalved, eared shells, resulting in the name 'fan' scallop being applied to some species. The ligament is small and triangular. Tiny tentacles along the edges of the mantle may bear light-sensitive 'eyes'. They may live free, cemented to the substrate or attached by a byssus. Some of the free-living species are capable of swimming by clapping the valves together.

SCALLOP *Pecten fumata*

Pecten shells are nearly spherical with a deeply convex right valve and nearly flat left valve. This species is sculptured with radial folds. Usually off-white with concentric bands of red-brown. There are different forms of this scallop but apparently only a single variable species. Grows to 120 mm diameter. Free-living on muddy or sandy bottom and fished commercially in southern Australia.

LEOPARD SCALLOP *Annachlamys flabellata*

Compressed shells with nearly equal ears. Sculpture of 20 radial ribs and fine concentric lamellae. Right valve white, left valve white with pink or purple ribs; both valves rose or pink inside. Grows to 100 mm diameter. Muddy sand in the subtidal zone. Northern Australia.

QUEEN FAN SCALLOP *Chlamys bifrons*

The fan scallops belonging to the genus *Chlamys* are characterised by unequal ears (the anterior ear longer than the other) and compressed and unequal valves, the right smaller and clasped by the left. This species usually has a lilac left valve and white right valve but colour is extremely variable. The valves are also differently sculptured, the right with five broad ribs, the left with narrower, more widely spaced ribs. Grows to 90 mm high. Sandy bottom in the subtidal zone. South-eastern Australia.

AUSTRALIAN FAN SCALLOP *Chlamys australis*

The largest of the Australian fan scallops and one of the most colourful. The colour range includes yellow, purple, brown and orange, but a heavy growth of sponges often obscures the full beauty of these shells. Sculptured with many multiple, prickly ribs. Grows to more than 110 mm high. Sponge beds on sandy bottom in the subtidal zone. Most often taken by trawlers. Central Western Australia.

DOUGHBOY FAN SCALLOP *Chlamys asperrimus*

Scaly sculpture of radial ribs. Variable colour. Grows to 90 mm high. Attaches by a byssus; usually covered by a layer of sponge. Common in subtidal seagrass beds. Southern Australia.

Scallop

Leopard Scallop

Australian Fan Scallop

Queen Fan Scallop

Doughboy Fan Scallop

ATKINS' FAN SCALLOP *Chlamys atkinos*

Delicate, very compressed, finely sculptured little shells with amazingly variable colours and patterns. Grows only to about 30 mm high. Attaches by a byssus under stones; very common washed up on southern beaches. Southern Australia.

LIVID FAN SCALLOP *Chlamys livida*

Tall and strongly sculptured with spiny radial ribs; those of the right valve are more numerous but lower than those on the left. The exterior may be brown, purple, yellow, orange or white. Grows to 70 mm high. Attaches by a byssus under stones. Northern Australia.

DRING'S FAN SCALLOP *Chlamys dringi*

Very compressed, with very unequal ears and broad radial ribs comprising several scaly ridges. Usually brown, orange or purple with white markings near the umbos. The left valve is paler coloured. Grows to 45 mm high. Attaches by a byssus under stones. Northern Australia.

SCALY FAN SCALLOP *Chlamys squamosa*

Very compressed; left valve almost flat. Unequal ears and scaly, low radial ribs. Very variable in colour — usually orange, yellow or purple. Grows to 55 mm high. Attaches by a byssus under stones and coral slabs. Widespread in the Indo-Pacific and common across northern Australia.

FRECKLED FAN SCALLOP *Chlamys lentiginosa*

Sculptured with many finely scaled ribs. Unequal ears. The right valve is slightly more convex than the left. Black or purplish brown exterior freckled with white spots; brown interior. Grows to 50 mm high. Very common attached under stones in the shallows across northern Australia.

BURIED FAN SCALLOP *Chlamys funebris*

Quite like the previous species but differently coloured. Usually red, orange or brown with zigzag white markings near the umbos; white interior. Grows to about 60 mm high. Common under stones in the intertidal and shallow subtidal zones of the north-west Australian coast.

RADULA SCALLOP *Semipallium radula*

Strong-shelled, rather flat, straight-sided, tall scallop. Sculptured with broad, rounded radial ribs. White valves; red-brown blotches on the ribs of the left valve. Grows to about 70 mm high. Attaches by a byssus to the underside of stones. Widespread in the Indo-Pacific and common across northern Australia.

TASMAN SCALLOP *Mesopeplum tasmanicum*

Fan-shaped and compressed, the right valve more convex than the left. Sculpture of compound radial ribs that are broad and flat on the right valve, narrow on the left. The valves are also differently coloured: the right valve is claret red, blotched or concentrically banded with white, and the left valve is much paler. Grows to about 40 mm high. Attaches by a byssus to stones. South-eastern Australia.

Atkins' Fan Scallop

Livid Fan Scallop

Dring's Fan Scallop

Scaly Fan Scallop

Buried Fan Scallop

Freckled Fan Scallop

Radula Scallop

Tasman Scallop

FIVE-RIBBED SCALLOP *Mesopeplum anguineum*
Small and solid, characterised by five very strong radial ribs, fine radial striae and an amazing range of external colour. Grows to about 40 mm high. Subtidal zone; rarely collected alive. Nevertheless the shells are moderately common on beaches across southern Australia.

ROYAL CLOAK SCALLOP *Gloripallium pallium*
Truly a glorious shell, notable for its strong, scaly ribs, big ears and bright colours. Usually red or purple with irregular white bands and bright orange around the inner margin. Grows to about 100 mm high. Attaches by a byssus to the undersides of stones and coral slabs. An Indo-Pacific scallop common across northern Australia.

THORNY OYSTERS Family SPONDYLIDAE

Thorny oysters are related to the scallops but their colourful shells are ornamented with lamellae or spines and they live cemented to the substrate by the right valve, which is usually more convex than the left. The short, strong hinge has prominent interlocking teeth and sockets and there is a deep, triangular ligamental pit. Most live in the tropics and there are many species in northern Australia. Only one species is known from the temperate waters of the south.

SOUTHERN THORNY OYSTER *Spondylus tenella*
Oval, equivalve shell. Sculpture of close radial ridges bearing profuse short and spatulate spines. Rosy red, orange, mauve or pink exterior; white interior. Grows to about 100 mm long. Common in the subtidal zone across southern Australia.

VICTORIA'S THORNY OYSTER *Spondylus victoriae*
Equivalve but higher than wide. About six primary ribs bearing long spines that are either simple or spatulate; numerous prickly ribs in between. White or pink exterior, sometimes orange at the umbos; white interior. Grows to 80 mm high. Widespread Indo-Pacific oyster that is common across northern Australia. (Also known as *S. wrightianus*.)

Five-ribbed Scallop

Royal Cloak Scallop

Royal Cloak Scallop

Southern Thorny Oyster

Victoria's Thorny Oyster

JEWEL BOX SHELLS Family CHAMIDAE

Jewel boxes are heavy shells which cement one of the valves to the hard substrate (the left valve in the genus *Chama*). The upper valve is convex with a rough surface that is usually strongly sculptured with spiny or leafy radial ribs. The hinge teeth are represented by long, curved interlocking ridges. Two of the three living cosmopolitan genera are recorded from Australia. *Chama* has at least eight Australian species.

LAZARUS JEWEL BOX *Chama lazarus*

Characterised by highly variable colour and strongly leafy lamellae on the radial ribs. Most specimens are white or lemon yellow with brown radial rays and red or yellow at the umbos. Grows to 100 mm diameter. Indo-Pacific and northern Australia.

BEAUTIFUL JEWEL BOX *Chama pulchella*

A smaller species reaching a diameter of only about 50 mm. Lower (attached) valve deeply concave; upper valve relatively flat. The upper valve has crenulate concentric lamellae and two radial ridges that may bear leafy fronds. White with two radial brown rays. Indo-Pacific and northern Australia.

BASKET CLAMS Family FIMBRIIDAE

There is only one living genus in this family, which is related to the lucines (see page 56). The two common Australian species have large, thick, heavy shells with strong concentric sculpture and strong hinge teeth. They have unusual anatomy, possessing a long excurrent siphon but no incurrent siphon.

COMMON BASKET CLAM *Fimbria fimbriata*

The shell outline is oval and the sculpture comprises low, slightly wavy concentric ribs and weak radials which become obsolete ventrally. White exterior, sometimes rose-tinted along the upper margin; white or yellow interior is light brown at the margin. Grows to about 95 mm long. Sand in the intertidal and shallow subtidal zones. Indo-Pacific and northern Australia.

SOWERBY'S BASKET CLAM *Fimbria sowerbyi*

Differs from the previous shell by its more quadrate outline and stronger concentric sculpture comprising thin, scalloped lamellae. Cream exterior with pink radial rays; white interior. Grows to about 100 mm long. Sand in the intertidal and shallow subtidal zones. Indo-Pacific and northern Australia.

Lazarus Jewel Box

Beautiful Jewel Box

Common Basket Clam

Sowerby's Basket Clam

LUCINES Family LUCINIDAE

There are many genera and species of this family in both high and low latitudes throughout the world. Lucine shells have a prominent, deeply impressed lunule and the hinge usually has two cardinal teeth in each valve and elongated lateral teeth. The pallial line is distinct and lacks a sinus. Lucines are burrowers in soft sediments. Some of them are found in environments with a high sulphide content and it seems likely that they supplement their diet through a symbiotic relationship with sulphur-oxidising bacteria held in their gills.

CORRUGATED LUCINE *Austriella corrugata*
Off-white shell, inflated and globular in outline, though slightly blunt posteriorly. Cardinal teeth are lacking in the right valve and represented by two nodules in the left valve; laterals are obsolete. The sculpture consists of widely spaced concentric lamellae. Grows to about 50 mm diameter. Common in mangroves across northern Australia, where it burrows in mud.

INTERRUPTED LUCINE *Codakia interrupta*
Circular with pointed umbos, short cardinal teeth and strong laterals. Sculpture with fine radial ribs and concentric ridges. Finely nodulose. White exterior; yellow interior and hinge. Grows to 40 mm diameter. Sand among coral rubble. Common across northern Australia.

RED-MARGINED LUCINE *Codakia paytenorum*
This abundant coral reef bivalve has a thick, strong, round shell that is laterally compressed. White exterior may be tinged with yellow or mauve. Yellow interior with a red margin. Sculpture of weak radial cords. Grows to about 60 mm diameter. Sand associated with coral reefs. Indo-Pacific and northern Australia.

TIGER LUCINE *Codakia tigerina*
Large compressed, circular shell with pointed umbos. Short cardinal teeth, strong laterals. Reticulate sculpture of about 50 radial ribs crossed by finely nodulose concentric cords. White exterior sometimes tinged with rose; yellow interior with rose-tinged margins; rose hinge. Grows to 100 mm diameter. Sand among rubble on coral reefs across northern Australia.

BEAUTIFUL LUCINE *Ctena bella*
Small, solid shell. White, yellow or pink exterior; white interior with a pale yellow tinge. Sculpture of rounded radial ribs crossed by fine concentric cords. Its height (25 mm) is greater than the length and the umbos are high and pointed. Sand. Common throughout the Indo-Pacific and northern Australia.

ORNATE LUCINE *Divaricella ornata*
This species is remarkable for its strongly sculptured shell bearing branching radial ribs which form pointed teeth around the margins. Small (to 25 mm diameter) white and round. Strong cardinals; lateral teeth are small or lacking. Sand. Indo-Pacific and northern Australia.

Corrugated Lucine

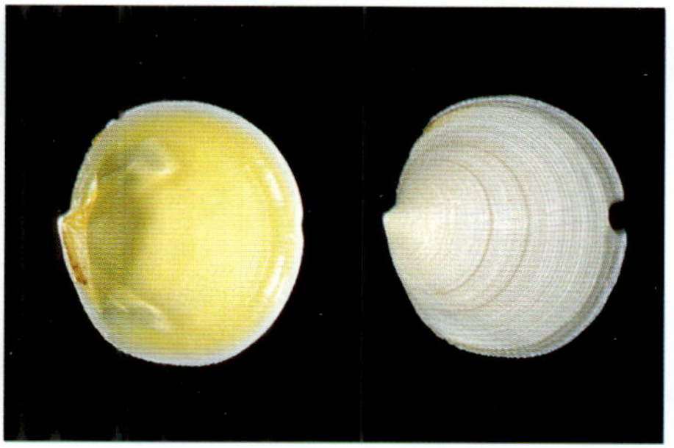

Interrupted Lucine

Red-margined Lucine

Tiger Lucine

Beautiful Lucine

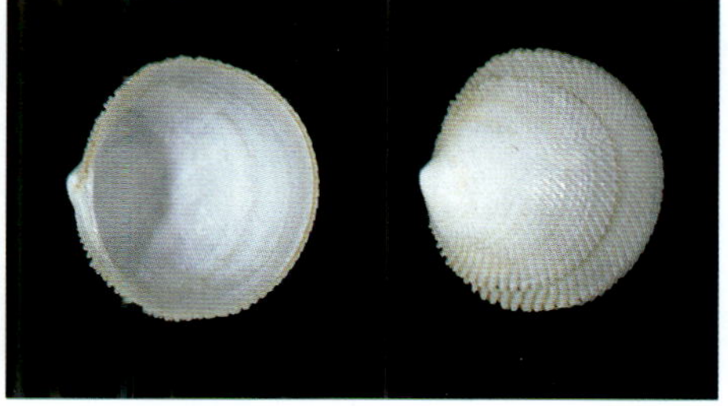

Ornate Lucine

CARDITAS Family CARDITIDAE

These strong, rounded shells have prominent umbos and a small, deep lunule. The anterior adductor muscle scar is on a raised platform, the pallial line is simple and there is a distinct pedal muscle scar. The curved hinge has two unequal cardinal teeth in each valve and long laterals. The sculpture is dominantly radial with crenulate margins. Some of the genera are deep-sea bivalves but three have species common in the shallows around Australian shores.

HALF ROUND CARDITA *Beguina semiorbiculata*
Largest of the family; grows to about 100 mm long. Laterally compressed, oval in outline though narrowing anteriorly, with terminal umbos and a sculpture of fine radial cords. Long, arched and strongly toothed hinge. Brown exterior, white anterior end; white interior but dark brown at the posterior margin. Attaches by a strong byssus within crevices of massive corals. Indo-Pacific and northern Australia.

AUSTRALIAN CARDITA *Cardita crassicosta*
Elongate–oval, narrowing anteriorly with terminal umbos and high radial ribs bearing prominent vaulted spines which become progressively larger posteriorly. Short hinge with small teeth. External colour extremely variable: usually white or cream, with brown crescent-shaped marks on the ribs, but brown, red and yellow specimens are common. White interior. Grows to about 50 mm long. Sand in the intertidal and subtidal zones. Circum-Australian.

STRONGLY RIBBED CARDITA *Cardita incrassata*
Oval, thick, heavy shell with subterminal umbos. Strong sculpture of about 16 rounded, nodulose radial ribs and strongly crenulate margins. Broad hinge with strong teeth. White, cream or pink with brown spots on the ribs; white interior. Grows to about 50 mm long. Sand in the intertidal and subtidal zones. Northern Australia.

MARBLED CARDITA *Cardita marmorea*
Very like the previous species but more elongate with pronounced umbos. The colour is more variable, ranging from fawn to red or orange with prominent brown blotches; white interior. Grows to 70 mm long. Abundant on intertidal flats across northern Australia.

PREISS'S CARDITA *Cardita preissii*
Similar to the two species above. Distinguished by its smaller size (up to 35 mm long) and more quadrate outline. Quite variable in colour, ranging from red to brown and orange, with brown spots; white interior. Common on sand flats across northern Australia.

PRICKLY CARDITA *Cardita muricata*
A small (to 30 mm long) elongate shell with a blunt anterior end and rather small, subterminal umbos. Many radial ribs, the posterior ones broad and bearing scales. Greyish cream exterior with red marks; white interior. An Indo-Pacific species that is common in northern Australia.

Half Round Cardita

Australian Cardita

Strongly Ribbed Cardita

Marbled Cardita

Preiss's Cardita

Prickly Cardita

HEART COCKLES Family CARDIIDAE

These shells are popularly known as heart cockles due to their heart-shaped outline in end view. They are distinguished by their hinge dentition: usually both valves have two cardinal teeth and distant laterals. There is no pallial sinus. The animals have short, tubular siphons which are fused together and surrounded by sensory papillae or tentacles. In some species the siphonal papillae bear minute 'eyes'. The large, sickle-shaped foot is capable of rapid digging. This large family has about 12 genera and 40 or more species burrowing in soft substrates around the Australian coast, most of them in the tropics.

WESTERN HEART COCKLE *Acrosterigma cygnorum*
This southern species of the genus is unusual in having a thin shell. White to lemon yellow exterior sculptured with fine radial cords, those at the posterior end being weakly spiny. Grows to about 40 mm high. Burrows in sand in the intertidal and shallow subtidal zones. Although named 'Western' because of its original locality, it is found across southern Australia.

REEVE'S HEART COCKLE *Acrosterigma reeveanum*
Like most members of the genus, this shell is solid and strongly sculptured with rough radial ribs and crenulate margins. Exterior cream below the umbos but otherwise orange with yellowish blotches. Grows to about 50 mm high. Common on intertidal flats across northern Australia.

UNICOLOURED HEART COCKLE *Acrosterigma unicolor*
Tall and tumid with a slightly blunt anterior margin. Sculpture of 45–50 close, rounded radial ribs. Most are lemon yellow but some are light brown and there may be brown blotches near the umbos; white or yellow interior. Grows to 40 mm high. Common on intertidal sand flats. Widely distributed in the Indo-Pacific and Queensland but not recorded from northern Western Australia.

STRAWBERRY HEART COCKLE *Fragum unedo*
Distinguished by its strong, wide shell with serrate margins and strawberry red spots on the strong ribs. Grows to 60 mm high. Another very widespread and common tropical species that lives in sand. Indo-Pacific and northern Australia.

SAW-TOOTHED HEART COCKLE *Fragum fragum*
Almost triangular with a straight, saw-toothed posterior side and rounded anterior. Sculpture of close, flat-topped ribs; anterior and posterior ribs nodulose. Cream or white with yellow flecks on the ribs. Grows to 35 mm high. Clean sandy lagoons and intertidal sand flats associated with coral reefs. Indo-Pacific and northern Australia.

Western Heart Cockle

Unicoloured Heart Cockle

Strawberry Heart Cockle

Reeve's Heart Cockle

Saw-toothed Heart Cockle

LYRATE HEART COCKLE *Nemocardium lyratum*

Sculptured with fine radial posterior ribs and sharp-edged, oblique ridges anteriorly. Polished outer surface is yellow but partly covered by a smooth, dark claret periostracum; white interior. Grows to 80 mm long. Sandy habitats. Indo-Pacific and northern Australia.

HALF HEART COCKLE *Lunulicardium hemicardium*

High, trapezoidal shells, sharply pointed ventrally and sharply keeled. Sculpture of low, flat radial ribs, the middle ones with a single row of small scales. White or cream. Grows to about 50 mm high. Very widespread and abundant on sand in the tropics. Indo-Pacific and northern Australia.

NARROWLY RIBBED HEART COCKLE *Fulvia tenuicostata*

This genus has thin, delicate shells with finely crenulate margins and a sculpture of fine, sharp-edged cords. The shells of this species are cream becoming yellow toward the margin with a hairy periostracum, the furry hairs growing along the tops of the radial ribs. Grows to about 50 mm diameter. Muddy sand across southern Australia.

THIN HEART COCKLE *Fulvia aperta*

Like the previous species but even thinner with a more flaring posterior end. Yellowish with purple flames and blotches, especially posteriorly and on the umbos. Grows to about 60 mm long. Sand in the subtidal zone. Common Indo-Pacific species found across northern Australia.

GIANT CLAMS Family TRIDACNIDAE

The so-called giant clams are tropical, shallow water bivalves. The largest species is *Tridacna gigas*, whose shell may weigh as much as 400 kg. The animals have fleshy, brightly coloured mantles in which are embedded symbiotic, photo-synthesising algae (zooxanthellae). There are also simple eyes in the mantle lobes. There is a strong byssus but this may atrophy in large individuals, which are so heavy that attachment to the substrate is unnecessary. Adult giant clams live a sessile existence on the sea floor in shallow water where there is ample light for the zooxanthellae. One species, the small *T. crocea*, is a borer in soft coral rocks. There are two genera totalling six species in Australian waters and they are all completely protected.

BEAR PAW CLAM *Hippopus hippopus*

The single living species of this genus has an inflated, subrhomboidal shell with long, squarish, marginal finger-like projections. Sculpture of about 13 spiny primary ribs. White to grey, usually with strawberry spots on the ribs. Grows to about 400 mm long. Unattached on sand and rubble in the intertidal and shallow subtidal zones. Indo-Pacific and northern Australia.

Lyrate Heart Cockle

Narrowly Ribbed Heart Cockles

Half Heart Cockle

Thin Heart Cockle

Bear Paw Clam

Bear Paw Clam (viewed from below)

GIANT CLAM *Tridacna gigas*

This is the famous Giant Clam, largest member of the family, which grows to 140 cm long. It has a very heavy, massive shell with four very prominent radial folds lacking concentric sculpture. The interlocking projections are long and pointed. Off-white. Unattached on sand or rubble in lagoons and channels. Indo-Pacific and northern Australia.

ELONGATE GIANT CLAM *Tridacna maxima*

Probably the most common and widespread species of the genus. Elongate, strongly inequilateral shell with a wide byssal opening and 6–12 fluted primary radial ribs forming long, pointed interlocking projections. White, sometimes yellow tinged. Grows to about 33 cm long. Rock and coral reefs, usually attached by the byssus in crevices on the reef surface. Indo-Pacific and northern Australia.

FLUTED GIANT CLAM *Tridacna squamosa*

A massive shell that reaches 400 mm in length. Distinguished by its semicircular, nearly equilateral outline, strongly fluted primary radial ribs and long, blunt interlocking projections. Off-white, sometimes with orange or yellow flutes. Coral reef species, usually unattached in sheltered areas. Indo-Pacific and northern Australia.

MACTRAS Family MACTRIDAE

Thin, smooth or concentrically sculptured shells with a smooth, rough or hairy periostracum. The family is characterised by an internal ligament set in a deep socket. The wide hinge plate has two cardinals in the right valve and a single, inverted v-shaped cardinal in the left valve. Lateral teeth are usually well developed. There is a prominent pallial line with a deep sinus. The animals have tubular siphons and a deep, hatchet-shaped foot. They are capable of rapid digging in sand or mud substrates.

SOUTHERN GAPER *Lutraria rhynchaena*

These differ from most in the family in having elongate–oval, strongly inequilateral shells that gape widely at the anterior end. The ligament is borne on a deep spoon-shaped part of the hinge. Sculpture of fine growth lines. Dull off-white exterior beneath an olive brown periostracum. Grows to 120 mm long. A widespread but uncommon sand-dweller in southern Australia.

AUSTRALIAN MACTRA *Mactra australis*

Triangular in outline with rounded anterior end and pointed posterior. Smooth surface (except for weak concentric ridges) white, suffused with blue and with a purple radial ray beneath a thin, straw-coloured periostracum. Grows to 40 mm long. Sand in the subtidal zone but commonly cast up on beaches in southern Australia.

Giant Clam

Giant Clam

Elongate Giant Clam

Fluted Giant Clam (juvenile)

Fluted Giant Clam

Southern Gaper

Australian Mactra

DISSIMILAR MACTRA *Mactra dissimilis*

Like the Australian Mactra but thin and sculptured with strong concentric ridges, especially posteriorly. Exterior is cream tinged with blue-grey and has blue umbone tips and a thin, yellow, shiny periostracum; bluish brown interior. Grows to about 50 mm long. Common along mainland beaches of northern and eastern Australia.

MACULATED MACTRA *Mactra maculata*

A thin-shelled mactra, triangular in outline with a smooth surface. Cream with light brown flecks and radial rays and purple-brown on the posterior area. The yellow-brown periostracum is thin and lamellate on the posterior area. Grows to 60 mm long. Sand on coral reef flats in the central Indo-Pacific and northern Australia.

TRUNCATE MACTRA *Mactra sericea*

Thin, oval and almost equilateral. Anterior end rounded, posterior slightly blunt. No sculpture except for low concentric ridges posteriorly. Fawn exterior with darker fawn radial rays and a violet patch on the umbos; white or fawn interior, sometimes violet posteriorly. Thin, smooth periostracum. Grows to 60 mm long. Sand in the intertidal and shallow subtidal zones of northern Australia.

CONTRARY MACTRA *Mactra contraria*

Moderately thick, triangular in outline. Anterior end narrowly rounded; posterior pointed and beaked. Fine sculpture of closely spaced concentric ridges that are wavy at the ends. Cream with a purple patch on the umbos. Grows to about 60 mm long. Sand. Eastern Australia.

REDDISH MACTRA *Mactra rufescens*

Triangular in outline like the Contrary Mactra and about the same size but both ends are pointed. Strong sculpture of concentric ridges. Cream with purple umbos. Sand. Commonly cast up on southern Australian beaches.

TRIGONAL MACTRA *Notospisula trigonella*

This solid little shell is highly variable in its shape, ranging from oval to elongate–oval with an angular, pointed posterior margin. Off-white to yellow exterior; white interior. Dense periostracum. Grows to about 20 mm long. Muddy sand in bays and estuaries and appears to be almost circum-Australian.

Dissimilar Mactra

Maculated Mactra

Truncate Mactra

Contrary Mactra

Reddish Mactra

Trigonal Mactra

TELLENS Family TELLINIDAE

Tellens have thin, laterally compressed shells that are inequilateral and inequivalve with a distinct posterior bend and a very deep pallial sinus. The hinge has two small cardinal teeth, but lateral teeth may or may not be present. These burrowers lie obliquely on their left valve beneath the surface of the sediment with their two long siphons thrust up into the water like snorkels. There are hundreds of species in this family and identification of most of them is very difficult. Some species are beautifully coloured and there are many colour patterns.

VIRGATE TELLEN *Tellina virgata*

Strong shell with an elongate–oval outline, rounded anterior, pointed and beaked posterior. Sculpture of close concentric lamellae. Typically cream with alternating radial rays of rose and yellow. White or yellow interior, but the external colour may show through. Grows to about 80 mm long. Very common on intertidal sand flats in the Indo-Pacific and northern Australia.

PERNA TELLEN *Tellina perna*

Thin, elongate shell with smooth, polished valves and a pointed, strongly beaked and flexed posterior end. White with a pale yellow or rose tinge; interior is often bright yellow. Grows to 80 mm long. Sand. A well-known Indo-Pacific bivalve that is common across northern Australia.

FOLIATED TELLEN *Tellina foliacea*

Thin and very compressed with rounded ends. Most of the surface is shiny with only minute microsculpture; posterior area of right valve is finely scaled. There is a row of marginal saw-tooth spines above the ligament. Dark yellow with a claret tinge on the interior. Grows to about 80 mm long. Sand. Indo-Pacific and northern Australia.

CROSS TELLEN *Tellina staurella*

Elongate–oval outline, narrowing posteriorly. Sculptured with close concentric lamellae. Yellow, but usually with two red rays radiating from the umbos (not visible in the illustrated specimen). Grows to about 50 mm long. Very common on tidal sand flats throughout the Indo-Pacific and northern Australia.

LITTLE WHITE TELLEN *Tellina albinella*

The name of this delicate tellen is misleading for it is typically uniformly pink or rose. It is also not so little; grows to about 60 mm long. Thin, smooth, shiny, elongate–oval shell, narrowing posteriorly. Sand. Commonly cast ashore on beaches of southern Australia.

WEDGE-SHAPED TELLEN *Tellina lilium*

This and related tellens have rather solid wedge-shaped shells with short posterior ends. Relatively small — less than 20 mm long. Usually white, but yellow near the umbos; bright yellow interior. Habitat not known but probably sand in the intertidal and shallow subtidal zones. Eastern Australia.

Virgate Tellen

Perna Tellen

Foliated Tellen

Cross Tellen

Little White Tellen

Wedge-shaped Tellen

RASP TELLEN *Tellina (Scutocopagia) scobinata*

Unlike most tellens this large species is round (lenticular in cross-section) with a diameter as much as 80 mm. The bend is pronounced in the right valve resulting in a distinct indentation in the margin, matched by a small projection in the left valve. Strong, scaly sculpture on both valves. Cream with faint brown radial rays. Sand-dweller on coral reefs throughout the Indo-Pacific and northern Australia.

DECUSSATED TELLEN *Pseudocopagia victoriae*

Rather plain, chalky white or grey bivalve. Round in outline, although the umbos are a little pointed and the dorsal margin angular. Grows to 55 mm diameter. Takes its common name from the finely decussate sculpture, that is, with intersecting radial and concentric striae. Sand under stones in the intertidal and shallow subtidal zones. Southern Australia.

SUNSET CLAMS Family PSAMMOBIIDAE

The bivalves of this family resemble the tellens, but lack a posterior bend and lateral teeth. The shells gape at both ends (especially posteriorly) and there is a deep pallial sinus. Like the tellens, they are deep burrowers and have two long siphons. There are four genera in Australia with 31 species. Most species live in the tropical north, but there are several in the temperate south.

LESSON'S GARI *Gari lessoni*

Compressed, broadly oval shell with an almost smooth surface. Cream to pale violet exterior with darker umbos; violet interior. There is a yellow-brown periostracum that is adherent around the margins. Grows to about 60 mm long. Muddy sand. Indo-Pacific and across northern Australia.

PACIFIC ASAPHIS *Asaphis violascens*

Strong, oval in outline, but slightly truncate posteriorly, and roughly sculptured with scaly radial ribs. Off-white or cream exterior, often with rose or pale purple radial rays; interior has a brilliant purple posterior end. Grows to about 60 mm long. Very common in sand and gravel under stones in the upper intertidal zone; often in rocky areas associated with mangroves. This species is collected for food in many parts of Asia. Widespread in the Indo-Pacific and across northern Australia.

DOUBLE-RAYED SUNSET CLAM *Soletellina biradiata*

Thin, compressed and oval with a smooth exterior and a thin, adherent periostracum. Exterior usually uniform yellow except for faint brown radial rays, slightly purple blotches near the umbos and two pale radial rays posteriorly which give the species its scientific and common names. Grows to 65 mm long. This deep-burrowing bivalve lives in sand. Very common on the beaches of sheltered bays and estuaries across southern Australia.

Rasp Tellen

Decussated Tellen

Lesson's Gari

Pacific Asaphis

Double-rayed Sunset Clam

SOLECURTUS CLAMS Family SOLECURTIDAE

These bivalves are remarkable for their very large bodies that cannot be accommodated within their shells. The shells tend to be elongate with blunt ends that gape widely but their hinge details are very like those of the sunset clams. They live in vertical burrows in sandy seabeds. There are two genera in Australia and about 10 species.

SULCATE SOLECURTUS *Solecurtus sulcatus*

These shells have parallel dorsal and ventral margins and rounded ends. Sculptured with concentric lamellae and narrow oblique ribs. Cream or rose coloured with four white radial rays. Grows to 80 mm long. Not common but widely distributed in the Indo-Pacific and across northern Australia.

WEDGE SHELLS Family DONACIDAE

This is a small but important family of agile bivalves that live on the lower slopes of sandy beaches. When uncovered by the waves they can quickly rebury themselves. Their shells are smooth, solid and triangular in outline — well fitted for rapid digging. They have two cardinal teeth in each valve and well-developed laterals. The ligament is external and there is a pallial sinus. There are seven Australian species. Perhaps the best known of them, and also the largest, is the Pipi, which is collected for bait and food along ocean beaches.

PIPI *Donax deltoides*

Also known as the Eugarie in Queensland and the Goolwa Cockle in South Australia. This is the largest of the Australian species; grows to 60 mm long. Triangular in outline with a rounded anterior end and angular posterior. Nearly smooth exterior may be brown, white, blue, purple, mauve, green, yellow or orange. Interior is usually purple. Ocean beaches in eastern and south-eastern Australia.

CUNEATE WEDGE SHELL *Donax cuneatus*

Almost oval and rounded at both ends. Smooth exterior, except for fine wrinkles on the posterior area. Like the Pipi, this species occurs in almost the entire colour spectrum but they are often rayed. Grows to about 35 mm long. Beaches in the western Pacific and northern Australia.

BRAZIER'S WEDGE SHELL *Donax brazieri*

Triangular in outline but usually elongate with the anterior end longer than the posterior. Minutely grooved exterior. Cream with two poorly defined darker radial rays. The growth lines are often well marked by a darker hue. Grows to about 20 mm long. Beaches in south-eastern Australia.

Sulcate Solecurtus

Pipi

Cuneate Wedge Shell

Brazier's Wedge Shell

PACIFIC BEAN WEDGE SHELL *Donax faba*

Almost oval with rounded ends and a steep posterior margin. There are concentric ridges on the posterior area but the remainder is smooth. Colour is extremely variable, covering the whole of the spectrum. Many specimens are strongly rayed. Grows to about 20 mm long. Beaches. Indo-Pacific and northern Australia.

VENUS CLAMS Family VENERIDAE

This is a very large family with many genera and at least 130 species in Australia, some of them abundant. In other parts of the world some venus clams are commercially harvested. Their shells are varied in form but characterised by three simple or bifid cardinal teeth in each valve. Anterior lateral teeth may also be present. The ligament rests in a groove along the dorsal margin and there is a moderately deep pallial sinus. Most species are burrowers. The siphons are short, so these bivalves live close to the surface of the sediment.

RETICULATE VENUS *Periglypta reticulata*

Large, solid, inflated shells. Slightly blunt posterior end. Strong radial and concentric sculpture forming nodules at the points of intersection. Cream exterior with light brown flecks; white interior, bright yellow or orange hinge. Grows to 90 mm long. Coral reefs, buried in sand beneath stones or coral slabs. Indo-Pacific and northern Australia.

YOUTHFUL VENUS *Periglypta puerpera*

Big and heavy like the previous species but more circular (less blunt posteriorly) and has finer sculpture with the concentric lamellae dominant. White with light brown radial rays that coalesce posteriorly. Grows to about 85 mm in diameter. Intertidal sand flats. Indo-Pacific and northern Australia.

CHEMNITZ'S VENUS *Antigona chemnitzii*

Large shell characterised by a dense sculpture of crimped concentric lamellae. Cream with light brown radial rays. Grows to about 100 mm in diameter. Sand in the intertidal and subtidal zones. Widespread across northern Australia.

PLANATELLA VENUS *Callista planatella*

Solid, oval in outline though narrowing posteriorly. The outside is shiny and strongly sculptured with concentric lamellae. Fawn exterior with crowded purple-brown radial rays; white interior. Grows to at least 90 mm long. Sand. Northern Western Australia.

TASMANIAN VENUS *Callista diemenensis*

Like its cousin above but smaller (to about 40 mm long) and almost smooth, sculptured only with fine concentric striae. Light brown exterior with darker brown radial rays. Buries in sand in the shallow subtidal zone. South-eastern Australia.

Pacific Bean Wedge Shell

Reticulate Venus

Youthful Venus

Chemnitz's Venus

Planatella Venus

Tasmanian Venus

WEDDING CAKE VENUS *Callanaitis disjecta*
A truly remarkable shell, oval in outline but with large, raised concentric frills from which its name is derived. Cream exterior tinged with pink and yellow; white interior. Grows to about 60 mm long. Sand. South-eastern Australia.

STEPPED VENUS *Katelysia scalarina*
The three species of the genus *Katelysia* in southern Australia are hard to tell apart. They all have strong shells with an elongate–oval outline and concentric sculpture. This one grows to about 40 mm long. The shells are slightly angular posteriorly. Strong concentric sculpture. Fawn exterior marked with light brown zigzag patterns; white or yellow interior, often purple at the muscle scars. Sand in bays and estuaries. Southern Australia.

RIDGED VENUS *Katelysia rhytiphora*
Largest of the three species of the genus, growing to 45 mm long. The outline is less angular posteriorly and the concentric lamellae are overlain by fine radial lirae. Cream or light brown exterior with a reticulate pattern of brown markings and sometimes radial rays. Character-istically the internal muscle scars are purple; sometimes the entire interior is purple. Very common in sand of bays and estuaries across southern Australia.

SULCOSE VENUS *Paphia crassiscula*
A strong, elongate shell with rather high umbos and strong, rounded concentric ridges. The outside is light brown with obscure purplish brown radial rays and zigzag patterns. Grows to 70 mm long. Sand in the intertidal and subtidal zones. Northern Australia.

LETTERED VENUS *Tapes literatus*
Thin for its size; grows to 110 mm long. Oval outline, slightly blunt at the rear end; laterally compressed. Sculpture of weak concentric ridges. White to yellow-brown exterior patterned with letter-like brown lines; interior is usually white but may be infused with yellow or rose. Sand. Common Indo-Pacific species widely distributed across northern Australia.

CHICKEN VENUS *Tawera gallinula*
Solid, oval shell, narrowing slightly posteriorly and sculptured with close concentric, faintly ribbed lamellae. Faintly toothed margin. Fawn exterior with thin, zigzag brown lines; interior usually violet near the margin and pink at the centre, sometimes entirely white. Grows to about 40 mm long. Sand. It is a mystery how this species came by its name. Southern Australia.

CAMP VENUS *Lioconcha castrensis*
This tropical genus has strong, oval shells with a striking reticulate pattern of concentric lines. This is the largest of the five Australian species and probably the best known. It has a thick, nearly round cream shell with a strong pattern of widely spaced zigzag lines. Grows to about 65 mm high. Sand associated with coral reefs. Indo-Pacific and northern Australia.

Wedding Cake Venus

Stepped Venus

Ridged Venus

Sulcose Venus

Lettered Venus

Chicken Venus

Camp Venus

SCRIPT VENUS *Circe scripta*

Australia has about nine species of this genus; they are very hard to identify. Their round shells are very compressed. This slightly angular species has a sculpture of concentric ridges. Colour variable, usually cream with brown radial rays; interior has blotches of purple or peach. Grows to 50 mm diameter. Intertidal sand flats. Very common in the Indo-Pacific and northern Australia.

PLACID VENUS *Placamen placidum*

This genus has thick shells with prominent umbos and very strong concentric sculpture. This temperate species has closely spaced concentric ridges. Cream with obscure fawn radial rays. Grows to 30 mm long. Common in intertidal and subtidal sand in eastern and south-eastern Australia.

HEAVY VENUS *Placamen gravescens*

Chunky with widely spaced concentric ridges and pointed umbos. Three wide brown radial rays on the cream exterior; interior strongly coloured with purple. Grows to about 30 mm long. Muddy intertidal sand. Central Indo-Pacific and north-western Australia.

SCALY VENUS *Anomolocardia squamosa*

Solid with a beaked posterior end. Sculptured with concentric and radial cords. Cream exterior; interior often has purple blotches at the ends. Grows to about 45 mm long. Abundant on muddy intertidal sand flats associated with mangroves. Northern Western Australia.

RIDGED DOSINIA *Dosinia scalaris*

This is a large genus of compressed lenticular shells with many Australian species. This is one of the largest, sculptured with concentric ridges that become lamellate at the extremities. White with brown radial rays; sometimes entirely brown. Grows to 60 mm diameter. Very common on intertidal sand flats across northern Australia.

BLUE-TINGED DOSINIA *Dosinia caerulea*

More inflated than most species of the genus. Finely sculptured with concentric lirae and faint radial striae. Usually white with coloured umbos but takes its common name from a faint bluish tinge present in many specimens. Grows to nearly 70 mm diameter. Commonly cast ashore on beaches in south-eastern Australia.

FAINTLY FRILLED VENUS *Bassina pachyphylla*

The shells in this genus have no lateral teeth. This solid species has an almost smooth shell. Fawn or light brown exterior; white interior. Grows to about 75 mm long Sand in the shallow subtidal zone. South-eastern Australia.

WHITE IRUS *Irus carditoides*

Shells of this genus tend to have irregular outlines but are generally quadrate and sculptured with prominent concentric lamellae. This white species has frilled lamellae. Grows to about 50 mm long. Intertidal sand flats. South-eastern Australia.

Script Venus

Placid Venus

Heavy Venus

Scaly Venus

Ridged Dosinia

Blue-tinged Dosinia

Faintly Frilled Venus

White Irus

Gastropods Class GASTROPODA

This is by far the largest class of the Mollusca. There has been a recent major re-assessment of gastropod classification (see Ponder, in Beesley et al. 1998, Part B) which recognises two subclasses. The traditional classification recognises three: prosobranchs, opisthobranchs and pulmonates. The arrangement in this book follows the traditional classification.

More than 200 families of these snails and slugs are represented in the Australian marine fauna. Thirty-nine of the families represented in shallow coastal waters are included in this book to illustrate the diversity of shelled gastropods commonly found along Australian shores.

Prosobranchs

This is a huge group with many families and countless species. Most of the species are shelled and many of them have an operculum that closes the shell aperture. Some terrestrial prosobranch snails have an operculum, which distinguishes them from their pulmonate cousins. Prosobranchs range from simple herbivorous species that graze on sea plants, to fish-eating predators.

LIMPETS Superfamily PATELLOIDEA

Limpets have simple, conical shells. Two of the five families in the group, the Acmaeidae and Patellidae, are conspicuous on Australian rocky shores where they browse on microalgae growing on the rocks. Their adductor muscle scars form a u-shape on the inner shell surface and enclose a spoon-shaped central area known as the spatula. Acmaeid limpets are mostly small and have a single gill in the mantle cavity. Patellid limpets have no gill; instead they have a row of respiratory filaments in shallow grooves along each side of the body. (Note: several unrelated gastropod families with simple conical shells are also known as 'limpets'.)

SUGAR ACMAEID LIMPET *Patelloida saccharina*
Elongate–oval, narrowing anteriorly. Sculptured with strong radial ribs. Greyish exterior, darker between the ribs; variably coloured spatula profusely peppered with brown spots. Grows to about 40 mm long. Middle and lower intertidal zones of rocky shores in exposed situations. Indo-Pacific and northern Australia.

TALL-RIBBED ACMAEID LIMPET *Patelloida alticostata*
An oval limpet with 12–30 strong radial ribs and a crenulate margin. White or brown exterior with crescent-shaped darker marks between the ribs; inner margin spotted with black; brown spatula. Grows to about 40 mm long. Rocks in the middle intertidal zone across southern Australia from the central east coast to the central west coast.

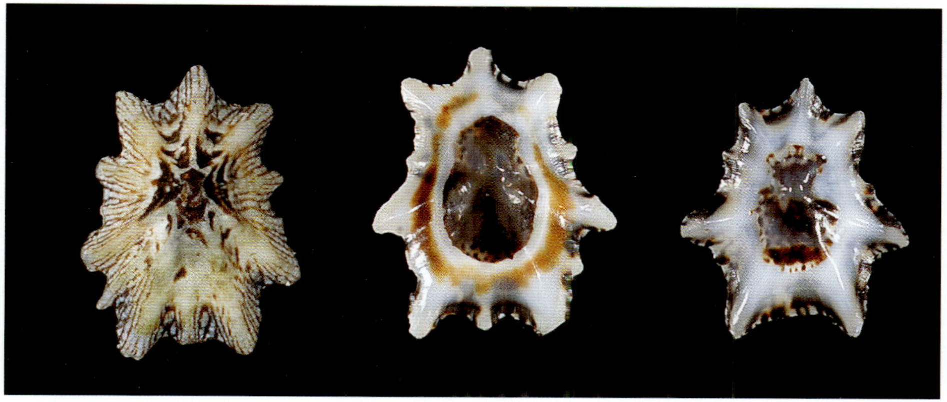

Sugar Acmaeid Limpet

Tall-ribbed Acmaeid Limpet

Limpet zone on a rocky shore with *Patella peronii* and *Notoacmea* sp.

RADIATE PATELLID LIMPET *Cellana radiata*

Roundly oval in outline with a low, subcentral apex, corrugated margin and sculpture of 11–14 primary ribs. Grows to about 40 mm long. Cream exterior with brown streaks or blotches. Interior has a brown or white spatula with brown edges, surrounded by a yellow zone; alternating white and brown blotches at the margin. Rocks in the upper and middle intertidal zone. Indo-Pacific and northern Australia.

SOLID PATELLID LIMPET *Cellana solida*

Oval in outline and sculptured with about 26 rounded radial ribs. Deeply scalloped margin. Dull grey to buff exterior is often rayed with brown. Interior has a bluish or slate grey spatula, yellow muscle scars and orange margin. Grows to 80 mm long. Conspicuous in the upper and middle intertidal zone of rocky shores in south-eastern Australia.

ARIEL PATELLID LIMPET *Cellana tramoserica*

Like the Solid Patellid Limpet and found in the same habitats, but usually smaller (up to 60 mm) with about 36 rather angular radial ribs and a finely crenulate margin. Colour is variable. Eastern and south-eastern Australia.

PERON'S PATELLID LIMPET *Patella peronii*

Oval with varied sculpture comprising about 16 or more primary radial ribs, fine threads in the interspaces and smooth or finely crenulate margin. Upper surface brown or grey with paler ribs; white interior often has a black margin, spatula yellow. Grows to about 50 mm long. Common in the upper intertidal zone, often forming a distinct limpet zone along rocky shores. Southern Australia.

STAR PATELLID LIMPET *Patella flexuosa*

This tropical limpet takes its common name from its radial ribs that project strongly at the margin. Narrowly oval outline, low apex. Grey to yellow exterior with brown blotches between the ribs; white interior with a yellow, orange-brown or grey spatula. Grows to 40 mm long. Rocky shores in the middle intertidal zone. Indo-Pacific and northern Australia.

MALTESE CROSS LIMPET *Patelloida insignis*

This small limpet is characterised by a prominent brown cross on the back. The inside is usually white with a brown spatula but the brown cross on the outside may show through. Grows to about 30 mm long. Under stones in the intertidal zone. Southern Australia.

NEGLECTED PATELLID LIMPET *Patella laticostata*

The largest Australian limpet; grows to 110 mm long. High profile, radial ribs, small teeth around the margin. Greenish grey exterior with brown rays; brown or fawn interior with brown muscle scars. Lower intertidal zone of rocky shores exposed to heavy wave action. Southern Western Australia.

Radiate Patellid Limpet

Solid Patellid Limpet

Ariel Patellid Limpet

Peron's Patellid Limpet

Star Patellid Limpet

Neglected Patellid Limpet

Maltese Cross Limpet

NERITES Family NERITIDAE

Most of these herbivorous marine, estuarine and freshwater snails are tropical — there is only one species in the temperate south of Australia (*Nerita atrementosa*). Marine nerites are gregarious, living in the middle or upper intertidal zone. They are thick-shelled and globular. The columellar lip is callused and wide, forming a flat shelf or 'deck' and there is a thick, tightly fitting, calcareous operculum with a peg-like basal tooth.

OX-PALATE NERITE *Nerita albicilla*
Almost planispiral with a small, low spire. Spirally corded exterior; finely granose operculum. Black or grey with fine, irregular flecks; deck green, yellow or white, heavily pustulose. Grows to about 25 mm wide. Middle intertidal zone on rocky shores. Indo-Pacific and northern Australia.

LINEATE NERITE *Nerita balteata*
The largest of the Australian nerites (grows to 40 mm high) and unusual in that it lives on the trunks and branches of mangrove trees. Globular and finely ribbed; smooth deck weakly toothed along the margin; granose operculum. Light brown with darker spiral lines; yellow or orange deck; grey operculum. Indo-Pacific and northern Australia.

PLICATE NERITE *Nerita plicata*
Globular and coarsely ribbed with a strongly toothed outer lip. Callused, wrinkled deck with four squared teeth on the margin; smooth operculum. White or yellow, sometimes spotted with grey; deep interior yellow; fawn operculum. Grows to about 30 mm high. High in the intertidal zone on rocky shores. Indo-Pacific and northern Australia.

ANCIENT NERITE *Nerita polita*
Shiny, smooth and globular with a flat spire and a smooth deck. The operculum is finely grooved near the margin. May be cream, brown, pink or green with irregular black or brown markings; orange aperture; red or dark orange deck; black, grey or fawn operculum. Grows to 30 mm high. Middle intertidal zone on rocky shores. Indo-Pacific and northern Australia.

WAVED NERITE *Nerita undata*
Globular with a high spire, finely threaded surface and finely toothed outer lip. Wrinkled deck with a toothed margin; operculum pustulose. Light brown, yellow, grey or purple-black; white aperture often stained with yellow; grey operculum. Grows to 40 mm high. Common tropical species found among oysters and on rocks in the middle and upper intertidal zone. Indo-Pacific and northern Australia.

CHAMAELEON NERITE *Nerita chamaeleon*
Takes its name from its extremely variable colouring. Granular surface is more or less grey or brown mottled or banded with many other colours; green and pustulose operculum. Grows to 20 mm wide. Rocky shores high in the intertidal zone. Indo-Pacific and northern Australia.

Ox-palate Nerite

Lineate Nerite

Plicate Nerite

Ancient Nerite

Waved Nerite

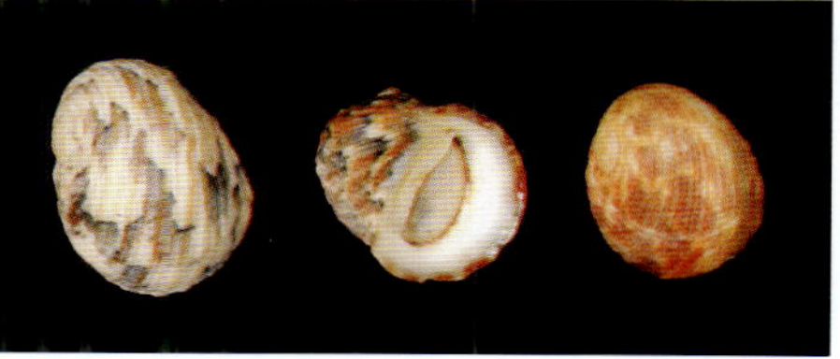

Chamaeleon Nerite

ABALONE Family HALIOTIDAE

Abalone are well known for their edible body and attractive shells which have brilliant mother-of-pearl on the inside. They are flattened, ear-shaped and asymmetrical with a low spire and a spiral row of open excurrent water holes around one side. These animals are herbivorous, feeding on algae. They adhere to rocks and have a wide, flat foot and powerful adductor muscles comprising the meat, which is considered such a delicacy. At one time they were called mutton-fish in Australia. There are about 19 Australian species. Some of the large gregarious temperate Australian abalone are fished commercially. In southern Australia in recent years several species have been successfully cultivated in tanks and provide a resource for a developing abalone aquaculture industry.

DONKEY'S EAR ABALONE *Haliotis asinina*
Unlike other abalone in that it has a thin, elongate shell and a fleshy mantle that almost covers the shell in life. Smooth, olive green exterior with irregular pale green or yellowish patches. Grows to 120 mm long. Common but not gregarious under stones on coral reefs and rocky shores. Indo-Pacific and northern Australia.

ROE'S ABALONE *Haliotis roei*
Oval with rough spiral cords of variable size. Reddish brown, sometimes with green rays. Grows to about 120 mm long. Abundant and gregarious on rocks in the lower intertidal and shallow subtidal zones. Fished commercially. South-western Australia.

SCALY ABALONE *Haliotis squamata*
Elongate with a rough, rounded exterior sculptured with finely beaded spiral cords and simple holes. Reddish brown with irregular greenish or cream zigzag rays and cream patches in a row around the outer margin. Grows to about 80 mm long. Common under stones in the shallow subtidal and intertidal zones but not gregarious. North-western Australia.

STAIRCASE ABALONE *Haliotis scalaris*
This is one of Australia's most spectacular shells. Oval in outline. Complexly sculptured with a thick, scaly central rib and many scaly spiral cords and thin radial lamellae between the rib and the spire, giving it the spiral staircase appearance. The holes are located on high conical tubercles. Orange-red with curved, radiating cream rays. Grows to 100 mm long. Common but never abundant under stones in the intertidal and subtidal zones. Southern Western Australia.

BLACK-LIP ABALONE *Haliotis ruber*
One of the larger species that is gregarious and fished commercially. Oval and roughly sculptured with spiral cords and broad radial folds. Reddish brown with greenish rays on the side. Grows to 160 mm long. Common on rocks in the lower intertidal and subtidal zones. Eastern and south-eastern Australia.

Donkey's Ear Abalone

Roe's Abalone

Scaly Abalone

Staircase Abalone

Black-lip Abalone

VARIABLE ABALONE *Haliotis varia*

Small, elongately oval with irregular radiating folds crossed by low, rounded, weakly nodulose spiral ribs. The holes are round and located on slightly elevated tubercles. Brown or greenish, usually with white or cream patches. Grows to 40 mm long. Common under stones in the intertidal and shallow subtidal zones. Indo-Pacific and northern Australia.

GREEN-LIP ABALONE *Haliotis laevigata*

An important commercial species. Quite large; grows to about 150 mm long. Rather smooth, greenish or reddish exterior, often rayed. The holes are simple. Rocks in the intertidal and subtidal zones. Southern Australia.

KEYHOLE AND SLIT LIMPETS Family FISSURELLIDAE

The shells of most members of this family are conical and limpet-like but they differ from the true limpets in having a hole at the apex or a slit in the rim. This fissure is functionally equivalent to the row of respiratory holes in abalone. Keyhole and slit limpet shells have a shiny porcelain-like inner surface. There is a horseshoe-shaped muscle scar which has its open side facing anteriorly. In some species the body is too large to fit beneath the shell so that the animal has a slug-like character. They graze on plant or animal tissues growing on firm substrates. There are about 12 Australian genera and over 50 species.

LATTICED KEYHOLE LIMPET *Diodora jukesii*

High, conical shell, oval in outline, roughly ribbed with the hole at the apex. Yellowish white, sometimes blotched with brown or with faint brown radial stripes. Grows to 55 mm long. Under stones. There are several Australian species of the genus and they are hard to tell apart. This one is from northern Australia.

SHIELD LIMPET *Scutus antipodes*

Also called the Boat Shell. This remarkable slug-like creature is a feature of the southern Australian coast. In life the fleshy black mantle covers the shell. Low, oblong, shield-like, off-white shell, sculptured with concentric striae. The fissure is represented by a shallow notch in the anterior margin. Grows to 110 mm long. Commonly found under stones in the shallow subtidal and intertidal zones. Southern Australia.

BLACK KEYHOLE LIMPET *Amblychilepas nigrita*

Saddle-shaped and oblong with an oval central hole. Sculptured with finely beaded radial cords. Most are cream with broad reddish brown rays. Grows to about 25 mm long. Under stones; the large, yellow, slug-like animal is very active. The species and common names seem to be inappropriate. Southern Australia.

Variable Abalone

Green-lip Abalone

Latticed Keyhole Limpet

Green-lip Abalone

Shield Limpet

Black Keyhole Limpet

JAVAN KEYHOLE LIMPET *Amblychilepas javanicus*
A misnomer because this limpet is found in the temperate waters of southern Australia — this locality error was common with species described by the Frenchman Lamarck. Squarish with sculpture of crowded concentric striae and brown radial rays. Grows to 25 mm long. Under stones around southern shores from southern Queensland to Fremantle in Western Australia.

RUGOSE SLIT LIMPET *Montfortula rugosa*
High, conical little shell with a distinct notch in the anterior margin representing the slit. Finely rugose (rough) and usually white with brown rays. The outline of the spatula on the inner side is prominently marked by dark green. Grows to about 20 mm long. Under stones. Southern Australia.

ELONGATED KEYHOLE LIMPET *Macroschisma producta*
Saddle-shaped and elongate–oblong with a long, triangular, posterior hole. Finely grooved, white or rose, usually with brown radial rays. The shell may grow to 20 mm long but the pale slug-like animal is much longer. Common under stones and in crevices in the intertidal zone of rocky shores. Southern Australia.

RIDGE-BACKED KEYHOLE LIMPET *Macroschisma munita*
Very like the previous species and about the same size but with a prominent angular ridge along the back. May be only a northern form of the same thing. Both occur in the same habitats in the south-west. West and north-west coasts of Western Australia.

ROUGH-BACKED KEYHOLE LIMPET *Macroschisma bakiei*
Broader and with a lower profile than the previous species. Roughly sculptured with nodulose ribs, the anterior ones stronger than the others. Pear-shaped hole. Usually rose, pink or pale brown. Grows to 28 mm long. Under stones in the intertidal and subtidal zones. Southern Australia.

TOP SHELLS Family TROCHIDAE

The common name of this very large family comes from the conical shape of many species, although some are turbinate or even ear-shaped. They have a thin, multispiral operculum made of horny material and mother-of-pearl on the inner surface. The animals are herbivores, carnivores or detritus feeders. Most live on hard substrates but some crawl on sand or mud. Some larger species are fished commercially for their meat and 'mother-of-pearl' shell. Only a small selection of the very many Australian species are illustrated here.

TROCHUS *Trochus niloticus*
This is the commercial trochus fished for its meat and mother-of-pearl shell. Largest species of the genus; grows to 150 mm high. Heavy, conical and weakly sculptured shell. The body whorl of mature specimens has concave sides and a thickened rim. Wide, funnel-like umbilicus. Off-white with oblique reddish stripes. Coral and rocky shores. Indo-Pacific and northern Australia.

Javan Keyhole Limpet

Rugose Slit Limpet

Elongated Keyhole Limpet

Ridge-backed Keyhole Limpet

Rough-backed Keyhole Limpet

Trochus

HANLEY'S TROCHUS *Trochus hanleyanus*

Flat-sided, angular at the periphery and sculptured with finely granose spiral cords and nodules. Cream or green sides with oblique red lines; the base has thin, wavy, crowded, radiating lines. Grows to about 50 mm high. Under stones in the intertidal and shallow subtidal zones. Very common throughout the Indo-Pacific including northern Australia.

WAVY TOP *Austrocochlea concamerata*

Turbinate shells encircled by about 12 spiral cords. Weakly lirate outer lip; one or two small nodules on the columella. Black or grey exterior with yellow or white spots on the ribs; white aperture. Grows to about 15 mm high. Common in the intertidal zone of rocky shores across southern Australia.

RIBBED TOP *Austrocochlea constricta*

Like the Wavy Top but taller with fewer ribs and a strongly lirate aperture. Usually grey or purplish black exterior with oblique creamy stripes or thin lines. Grows to about 25 mm high. Rocky shores and muddy flats in sheltered bays and estuaries. Southern Australia.

ONE-TOOTHED TOP *Monodonta labio*

Turbinate shell with weakly granulated spiral ribs, lirate aperture and arched columella with one strong basal tooth. Grey, green, brown or red with darker spots on the ribs. Grows to about 40 mm high. Abundant high in the intertidal zone of rocky shores in the tropics. Indo-Pacific and northern Australia.

PEARLED TOP *Herpetopoma aspersa*

Globular little shells with a high spire and beaded spiral ribs. There is a tiny tooth at the base of the columella. The silvery internal mother-of-pearl gives the species its common name. Grows to about 15 mm high. Very common under stones on rocky shores across the southern Australia and well up both the east and west coasts.

KEELED CLANCULUS *Clanculus limbatus*

Conical like the other members of this genus, with a very nodulose columella and toothed lip. Sculptured with nodulose spiral ribs and a prominent peripheral keel. Fawn with brown markings. Grows to about 15 mm high. Very common under stones in the intertidal and subtidal zones of rocky shores. Southern Australia.

ROUNDED CLANCULUS *Clanculus maxillatus*

Rather like the previous species but smaller and lacks a peripheral keel. Pale brown or grey with dark brown and white nodules on the ribs. Grows to 11 mm high. Very common under stones on rocky shores from western South Australia to Kalbarri on the central west coast of Western Australia.

Hanley's Trochus

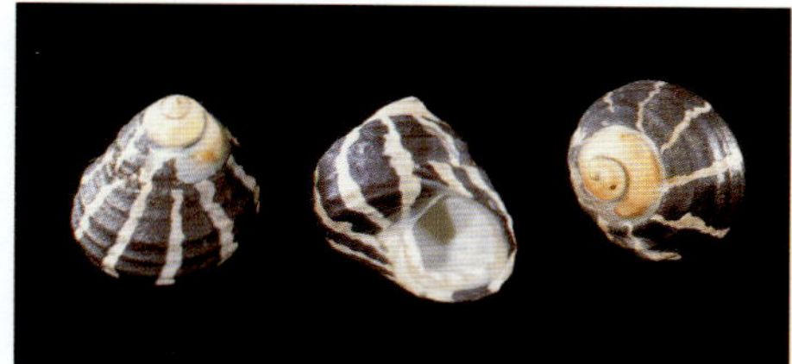

Ribbed Top

Pearled Top

Rounded Clanculus

Wavy Top

One-tocthed Top

Keeled Clanculus

MAUGER'S CLANCULUS *Clanculus maugeri*

One of the larger species in the genus; grows to 20 mm high. Conical rather than turbinate and rather straight-sided with an angular peripheral keel. Reddish brown, finely spotted with white. Under stones in the subtidal zone. South-eastern Australia.

KELP SHELL *Phasianotrochus eximius*

Rather thin-shelled top with a tall spire (up to 40 mm high) and a smooth, glossy surface encircled by incised lines. Lirate aperture, smooth columella and weak basal tooth. Fawn, olive green, brown or rose exterior; brightly iridescent interior. Among kelp; commonly cast ashore after storms. Southern Australia.

NECKLACE KELP SHELL *Phasianotrochus bellulus*

Unlike the previous species, this little shell is stout and solid with a rounded periphery. Varicoloured glossy exterior, usually red with oblique bands outlined with white; iridescent interior. Grows to 18 mm high. Common in beach drift. Widely distributed across southern Australia as far north as Kalbarri in the west but apparently not present on the east coast.

CONICAL TOP *Thalotia conica*

One of the most abundant tops in southern Australia. Tall, conical and sculptured with finely granose spiral ribs. Reddish brown exterior, darker nodules; lustrous silver interior. Grows to 25 mm high. Seagrass beds; sometimes washed ashore in vast numbers. Southern Australia from Bass Strait to Geraldton.

PRETTY TOP *Thalotia chlorostoma*

Possibly not closely related to the former species in spite of its generic classification. Conical with straight sides and angular peripheral keel. Extremely variable in colour — most commonly red, sometimes green, with a variegated pattern of spots and dashes. Brightly iridescent interior. Grows to 30 mm high. Algal beds in the intertidal and subtidal zones. South-western Australia.

IMBRICATE PEARL SHELL *Granata imbricata*

This depressed shell has a very unusual form for a top shell, being ear-shaped like an abalone but lacking respiratory holes. It has a horny operculum like other tops. Finely sculptured cream exterior. It doesn't make pearls — the common name comes from its silvery internal mother-of-pearl. Grows to about 30 mm wide. Under stones in the intertidal and subtidal zones. Widely distributed across southern Australia from the central east coast to the central west coast.

SWOLLEN STOMATELLA *Stomatia phymotis*

Another ear-shaped top shell but not closely related to the previous species. The stomatellas have a twisted spire. This species has a rough exterior and is very variable in shape. Usually grey to cream exterior with brown spiral bands and spiral lines of red dashes. Brightly iridescent interior. Grows to 30 mm wide. Under stones in the intertidal and shallow subtidal zones. Indo-Pacific and northern Australia but extending around the south-west corner onto the south coast.

Mauger's Clanculus

Necklace Kelp Shell

Conical Top

Imbricate Pearl Shell

Kelp Shell

Pretty Top

Swollen Stomatella

TURBAN SHELLS AND THEIR RELATIVES
Family TURBINIDAE

There are several distinctive subfamilies in this group of herbivorous snails, including the true turbans, pheasant shells and dolphin shells. Turbans are thick, few-whorled, turbinate or conical shells with mother-of-pearl on the interior and a thick calcified operculum. Pheasant shells are thinner, glossy and very colourful and also have a thick, calcified operculum. Dolphin shells (not considered here) have a horny operculum like that of the top shells. The meat of large turbans is eaten in some parts of the world and shells of some Australian species are common in Aboriginal 'kitchen middens'.

PAINTED LADY PHEASANT SHELL *Phasianella australis*
This is the largest of the pheasant shells (to 80 mm high) and among the most spectacular of all Australian shells. Turbinate, tall-spired, rather thin with a glossy, brightly coloured exterior with a variety of complex patterns. Oval operculum is pointed posteriorly. Seagrass beds in the subtidal zone. Southern Australia from Bass Strait to Geraldton.

SWOLLEN PHEASANT SHELL *Phasianella ventricosa*
Similar to the Painted Lady, but shorter and more tumid. Brightly coloured exterior with a huge variety of patterns. Grows to about 40 mm high. Seagrass beds and among macroalgae of rocky shores. Southern Australia from New South Wales to the central west coast.

CAT'S EYE TURBAN *Turbo petholatus*
Turbinate and solid with rounded whorls. The smooth, polished exterior is usually brown, red or green with darker axial and spiral lines, the columella yellow, orange or green. The operculum has a green centre and brown margin. The common name comes from the round, eye-like operculum, which is commonly used in shell jewellery. Grows to about 60 mm high. Under stones on coral reefs and rocky shores. A well-known Indo-Pacific species common across northern Australia.

WAVY TURBAN *Turbo undulatus*
Solid and globular with rounded whorls which may bear weak spiral ribs. Oval operculum with a central hump encircled by a shallow channel. Pale green exterior with dark green wavy axial lines. Grows to about 60 mm wide. Among weed-covered rocks in the subtidal and intertidal zones. South-eastern and southern Australia but not on the west coast.

HEAVY TURBAN *Turbo torquatus*
Large, common shell with rounded (eastern specimens) or keeled (western specimens) whorls sculptured with spiral ribs and axial lamellae. Prickly, oval operculum with two thick spiral ribs. Sand-coloured, sometimes with green or orange mottling. Grows to 110 mm high. Among weed-covered rocks in the subtidal and intertidal zones. Southern Australia and well north on both the east and west coasts.

Swollen Pheasant Shell

Cat's Eye Turban

Painted Lady Pheasant Shell

Wavy Turban

Heavy Turban

MOON TURBAN *Turbo cinerea*
Smaller than most of the other turbans; grows to about 35 mm wide. Thick shell with a low spire and a sculpture of finely nodulose spiral cords. There is a curious spout-like projection at the base of the columella. Varicoloured. Operculum finely granose, white with a dark green outer band. Among rocks in the intertidal zone. Indo-Pacific and northern Australia.

SILVER-MOUTH TURBAN *Turbo argyrostomus*
Heavy with rounded whorls and a tall spire. Strongly sculptured with spiny spiral ribs. Light brown or grey with green and dark brown spots on the ribs and spines. Silvery interior. Thick, coarsely granose, mottled operculum. Grows to 80 mm high. Under and among stones and corals in the intertidal zone. Widespread on coral reefs in the Indo-Pacific and across northern Australia.

BEAUTIFUL TURBAN *Turbo pulcher*
Like the previous species in shape and about the same size but lacks spines and the operculum is very heavily granose. Fawn exterior with wavy green-brown axial stripes and brown blotches on the ribs; white operculum. Among rocks and corals in the intertidal and shallow subtidal zones. South-western Australia.

SCALY STAR SHELL *Astralium squamiferum*
One of several star shells, this species has a low spire and sharply keeled peripheral flange. The sides of the whorls are flat, granose and fluted above the keel. The operculum is oblong and concave at the centre. Exterior usually yellow or grey, but sometimes pink as in the illustrated specimen. Grows to about 30 mm wide. Very common in seagrass beds across southern Australia.

CREEPERS Family CERITHIIDAE

The creepers are a diverse group with very many species, mostly in the tropics. The long, tapering, many-whorled shells have an upturned anterior canal and a thin, subcircular, horny operculum. These gregarious detrital feeders live on sandy or muddy substrates and leave winding trails in the sediment as they wander about — hence the common name for the family.

SPINY CREEPER *Cerithium echinatum*
Stout shell, usually with a central row of prominent spiny nodules and knobby cords around the whorls. Short anterior canal is partly covered by the lip. Cream with purplish brown spots; white aperture. Grows to about 80 mm long. Crawls about on sand among rocks and coral rubble. Very common and widespread in the Indo-Pacific and northern Australia.

Moon Turban

Silver-mouth Turban

Beautiful Turban

Scaly Star Shell

Spiny Creeper

GIANT NODULOSE CREEPER *Cerithium nodulosum*
One of the largest living species of the family; grows to 150 mm long. Heavy shell bearing
prominent tubercles and a number of weakly nodulose spiral ribs. White exterior with brown
blotches but usually thickly covered with calcareous growths; white interior. Common among
rubble on coral reefs of the Indo-Pacific and northern Australia.

ALUCO CREEPER *Pseudovertagus aluco*
Large, solid species with a sharply upturned and elongated anterior canal partly covered by the
lip. Relatively smooth, white or cream whorls are heavily blotched and finely spotted with
purplish brown. Grows to 90 mm long. Sand in shallow pools on tidal flats. Central Indo-Pacific
and northern Australia.

BANDED CREEPER *Rhinoclavis fasciata*
Slender and acutely tapering shell. High, nearly vertical anterior canal. The early whorls are faintly
axially ribbed but the later whorls smooth and glossy. Colour and pattern are variable — usually
white with brown bands and spiral lines or rows of spots. Grows to more than 90 mm long.
Common in colonies on sandy flats and in shallow lagoons. Indo-Pacific and northern Australia.

COMMON CREEPER *Rhinoclavis vertagus*
A solid shell with tumid whorls that are smooth except for axial ribs near the suture. Early whorls
are varicose. Long, almost vertical siphon canal. Cream, white, yellow or orange, sometimes with a
tan spiral band behind the suture. Grows to about 70 mm long. Extremely abundant in sand on
tidal flats and shallow lagoons. Ubiquitous in the central Indo-Pacific and northern Australia.

DOUBLE-BANDED CREEPER *Clypeomorus bifasciata*
There are several species of this genus on muddy tropical intertidal flats and they are hard to
tell apart. This species has beaded spiral cords and the last varix on the ventrally flattened body
whorl is opposite the lip. Usually grey with black nodules and brown blotches. Some specimens
have two spiral bands. Grows to 30 mm long. Indo-Pacific and across northern Australia.

TOWER CREEPERS Family CAMPANILIDAE

This family has only a single living species, which is a relic of the family which flourished in
tropical seas of the world millions of years ago.

LIGHTHOUSE TOWER *Campanile symbolicum*
Very large shell; grows to at least 200 mm long. Flat sides and almost no sculpture except for
weak growth lines and a low spiral ridge just behind the suture. Short, nearly horizontal
anterior canal. Chalky white. Sand among weed-covered rocks. South-western Australia.

Giant Nodulose Creeper

Aluco Creeper

Banded Creeper

Common Creeper

Double-banded Creeper

Lighthouse Tower

MUD CREEPERS Families POTAMIDIDAE and BATILLARIIDAE

Mud creepers are similar to the creepers (see page 98). The shells have short anterior canals and the lower lip projects beyond the columella base. The operculum is thin, circular, multispiral and horny. They are intertidal snails, living in muddy habitats. There are two families, the tropical Potamididae and the southern Australian Batillariidae. There are not many species in either family but they may be very conspicuous in vast numbers in mangroves or salt marshes where they graze on organic detritus or microalgae. Like the creepers, they leave winding trails in the mud.

TELESCOPE MUD CREEPER *Telescopium telescopium*
Long, thick and conical with straight sides and a broad, flat base and no varices. Short whorls sculptured with thin spiral grooves. Brown, often with a single spiral central white line on each whorl; yellow columella. Grows to 110 mm long. Widespread in mud among mangroves of the Indo-Pacific and northern Australia.

GIANT MUD CREEPER *Terebralia palustris*
Heavy shell with a flat-sided spire, deeply incised suture and rounded base. Whorls have low varices (axial folds) and deeply incised spiral grooves. Brown with a fawn columella and purple-brown lip. Grows to 190 mm long. Large colonies on mud in and near mangroves in the Indo-Pacific and northern Australia.

STRIATE MUD CREEPER *Terebralia semistriatus*
Solid and pendant-shaped with deeply incised suture and inflated, spirally grooved whorls. Distinguished by its widely flaring outer lip, which covers the anterior canal. Brown with shiny brown and whitish patches on the columella and outer lip. Smaller than the Giant Mud Creeper; grows to about 50 mm long. Abundant in the mud among mangroves in north-western Australia. A better known and more widely distributed species in northern Australia and the Indo-Pacific region is *T. sulcatus*.

MANGROVE MUD CREEPER *Cerithidea anticipata*
Tropical, mangrove-dwelling mud creeper with a long, thin shell. Strong sculpture of nodulose axial ribs and spiral cords. Outer lip smooth but flaring. Fawn to brown, sometimes with darker spiral bands. Grows to about 45 mm long. Northern Australia.

HERCULES CLUB *Pyrazus ebeninus*
Tall-spired with knobby, spirally corded and axially ribbed whorls. Both the outer lip and the columella are callused and outward flaring. Dark brown exterior; purple in the deep interior but creamy on the lip and columella. Grows to about 110 mm long. Estuarine mud flats along the east coast of Australia.

Telescope Mud Creeper

Giant Mud Creeper

Striate Mud Creeper

A colony of Striate Mud Creepers on muddy sand of the upper intertidal zone of Shark Bay, Western Australia

Mangrove Mud Creeper

Hercules Club

SOUTHERN MUD CREEPER *Velacumantus australis*
Tall-spired with rounded whorls strongly sculptured with axial folds and spiral cords. The outer lip is simple. Dark brown or grey exterior, sometimes with a white central band; dark brown interior. Grows to 45 mm long. Vast colonies in sandy areas of estuaries and bays among seagrass and algae. Eastern and southern Australia.

WINKLES Family LITTORINIDAE

Winkles are small herbivorous snails with turbinate or conical shells. They live gregariously high in the intertidal zone where they are covered by water only during periods of high tide. They are adapted to frequent exposure to sun and air. Some actually live above the high-tide level where they are wet only by wave splash. They are conspicuous on rocky shores almost everywhere and graze on microalgae growing on the substrate. A few species live in tidal marshes or mangroves. There is a group of tropical, thin-shelled species that live on mangrove trees.

STRIPED-MOUTH CONNIWINK *Bembicium nanum*
Conniwinks are flat-based, conical snails that live on rocks in the intertidal zone. This one is cream to brown with grey mottling or wavy stripes on the sides and a cream or orange columella. Grows to about 12 mm wide. Rocky shores. Eastern and southern Australia.

BANDED PERIWINKLE *Littorina unifasciata*
Tall-spired, turbinate shell with convex whorls that are slightly keeled on the base and a sculpture of weak spiral striae. White or grey exterior with a central pale blue band; brown interior and columella. Grows to about 20 mm high. Large colonies high in the intertidal zone on rocky shores. Southern Australia, New Zealand and islands of the south-eastern Pacific.

UNDULATE PERIWINKLE *Littoraria undulata*
Tall-spired, turbinate shell with a rounded periphery and minute sculpture of spiral striae. Usually yellowish grey to brown exterior, banded or with brown wavy lines or spots; yellow-brown interior, violet columella. Grows to 20 mm high. Rocky shores. Indo-Pacific and northern Australia.

SCABRA PERIWINKLE *Littoraria scabra*
This is the largest of the thin-shelled, tall-spired, turbinate species that live on mangrove trees. Keeled around the base and sculptured with numerous low ribs. Pale grey with black or dark brown dashes forming oblique stripes. Grows to about 35 mm high. Trunks and roots at the seaward edge of mangroves throughout the Indo-Pacific and across northern Australia.

THIN PERIWINKLE *Littoraria filosa*
Similar shape to the previous species but thin-shelled and sculptured with prominent spiral ribs. Varicoloured but usually shades of yellow. Grows to about 35 mm high. Mangrove trees. Central Indo-Pacific and across northern Australia.

Southern Mud Creeper

Banded Periwinkle

Scabra Periwinkle

Thin Periwinkle

Striped-mouth Conniwink

Undulate Periwinkle

A colony of Scabra Periwinkles on a
mangrove trunk in the upper intertidal zone

PYRAMID NODIWINK *Nodilittorina pyramidalis*

Tall-spired, small, conical shell with a strong sculpture of granose or nodulose spiral cords. Blue-grey exterior with fawn nodules; brown interior and columella. Grows to only 15 mm high. Great numbers high in the intertidal zone of rocky shores. Indo-Pacific and northern Australia.

AUSTRALIAN NODIWINK *Nodilittorina australis*

Turbinate with a relatively low spire and a sculpture of fine spiral striae and low axial folds. Yellow-grey exterior; violet to tan columella. Grows to about 20 mm high. High on rocky shores. A Western Australian species found from Esperance to the Kimberley.

STROMBS Family STROMBIDAE

The strombs are tropical, herbivorous snails that live on sandy or rubble substrates in the intertidal and shallow subtidal zones. They have large eyes on long stalks that may be swivelled about. There is usually a flaring or thickened outer lip with a deep u-shaped notch near the anterior end that is used as a peep hole for the right eye when the animal is lying flat on the seabed. An elongate, horny operculum has a saw-toothed edge and is used as a lever when the animal is crawling about with its characteristic jerking movement. In the genus *Lambis*, which has several Australian species, the outer lip bears long spines (measurements are taken from the tips of the spines). At the other extreme *Terebellum* is smooth, cylindrical and the simple lip is spineless.

SPIDER CONCH *Lambis lambis*

Large, solid shells with heavy shoulder nodules. Flaring outer lip with seven straight or slightly curved spines. White or cream exterior with brown or bluish patches; glossy pink interior with a smooth, orange or purple-tan columella. Grows to 200 mm long. A very common creature on coral reefs. Indo-Pacific and northern Australia.

CHIRAGRA CONCH *Lambis chiragra*

Like the Spider Conch but larger. Flat-based with six thick, curved spines and a strongly lirate aperture. White exterior with brown patches and flecks; rose pink or orange interior with a smooth, shiny columella that is orange-violet or cream between the white lirae. Grows to 250 mm long. Among rubble on coral reefs. Indo-Pacific and northern Australia.

SCORPION CONCH *Lambis scorpius*

Seven spines like the Spider Conch but the central ones are flattened, nodulose and bent backwards at their ends. Cream to grey exterior with brown flecks; purple aperture with white lirae, surrounded by shiny orange on the base. Grows to about 170 mm long. Another coral reef conch but not as common as the others. Widespread in the Indo-Pacific but in Australia it has been found only in northern Queensland.

Pyramid Nodiwink

Australian Nodiwink

Spider Conch

Chiragra Conch

Scorpion Conch

SILVER STROMB *Strombus lentiginosus*

Very heavy, nodulose species with a thick but spineless lip. Off-white or grey exterior mottled with green or brown and bearing chestnut spots at the shoulders; often glazed on the underside of the body whorl. Smooth aperture is orange becoming cream near the margin. Grows to 100 mm long. Shallow water around coral reefs. Widespread in the Indo-Pacific and northern Australia.

DIANA STROMB *Strombus aurisdianae*

This shell has a tall, conical, nodulose spire and a flaring outer lip bearing a long, backward-pointing spike at the rear. The aperture is smooth and shiny. Cream exterior mottled with brown and blue-grey; rich orange-brown interior, thickly glazed on the columella. Grows to about 80 mm long. Pools and lagoons on coral reefs. Indo-Pacific and north Queensland.

VOMER STROMB *Strombus vomer*

Similar to the Diana Stromb and about the same size but with a white, strongly lirate aperture and stronger nodulose sculpture. Intertidal sand flats. Indo-Pacific and north-western Australia.

MUTABLE STROMB *Strombus mutabilis*

Relatively small with broad shoulders, a low, conical spire, simple outer lip and strongly lirate columella. Exterior colour is variable, usually cream with patches of orange, brown or yellow; pink interior. Grows to about 40 mm long. Common in the intertidal zone on rocky shores and coral reefs. Very widespread in the Indo-Pacific and northern Australia, extending well into the temperate zone on both the east and west coasts.

STRAWBERRY STROMB *Strombus luhuanus*

Unusual with its inturned outer lip and obconical outline resembling that of a cone shell. Smooth, cream exterior with pale brown flecks beneath a thick green-brown periostracum; rich orange or red interior with a black-brown columella. Grows to about 70 mm long. Colonies on grassy intertidal flats on coral reefs. Central Indo-Pacific and northern Australia.

CAMPBELL'S STROMB *Strombus campbelli*

This shell has a very tall, pointed spire, widely flaring outer lip and more or less smooth, white aperture. The exterior is usually off-white with a dense pattern of brown spiral and axial lines but pink and mauve forms occur. Grows to about 70 mm long. Intertidal sand flats and shallow sandy lagoons across northern Australia.

GIBBOSE STROMB *Strombus gibberulus*

Oddly asymmetrical, the spineless outer lip of this species is slightly thickened and flared. Smooth, white exterior with spiral bands of brown blotches; lirate interior is either white or tinted violet, brown or yellow with an elongate brown patch. Grows to about 60 mm long. Colonies on grassy intertidal flats. Indo-Pacific and northern Australia.

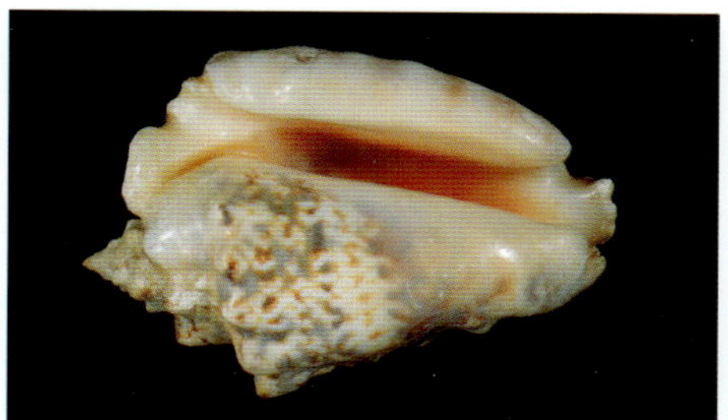

Silver Stromb

Vomer Stromb

Mutable Stromb

Campbell's Stromb

Diana Stromb

Strawberry Stromb

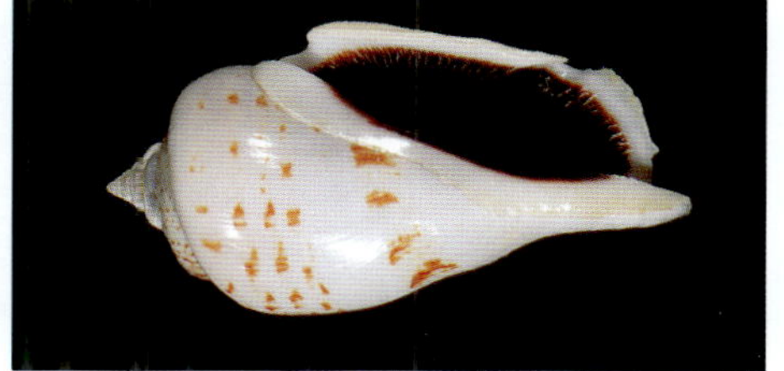

Gibbose Stromb

LITTLE BEAR STROMB *Strombus urceus*

Tall, nodulose spire, axial folds on the body whorl and a lirate aperture. Colour is very variable – exterior may be brown, off-white or grey with spiral lines and blotches of various hues. The aperture is usually dark purple or black, often becoming yellow on the columella. Grows to about 50 mm long. Muddy intertidal sand flats in the central Indo-Pacific and northern Australia.

TEREBELLUM CONCH *Terebellum terebellum*

Elongate and cylindrical (bullet-shaped) with a long, narrow aperture that is widely open at the anterior end. Smooth, glossy cream exterior may be spotted with red-brown or bear zigzag axial lines. Grows to 70 mm long. Sand in the intertidal and subtidal zones. Indo-Pacific and northern Australia.

COWRIES Family Cypraeidae

The cowries have oval, glossy, beautifully coloured and patterned shells characterised by prominent ridges or 'teeth' along the lips on both sides of the narrow aperture. The high polish of the cowry shell is produced by secretory cells in the skin-like, retractable mantle that spreads over its surface. Cowry shells are favoured everywhere for decorative purposes. There are more than 80 species in Australian waters including a dozen or so that live in the temperate waters of the southern coasts. Most are generalised browsers that live in rocky or coral reef habitats. During the day most of them hide under stones, emerging to go about their affairs at night. The species chosen here to represent the family are common in the intertidal and shallow subtidal zones.

MILK-SPOT COWRY *Cypraea vitellus*

Tumid and oval with rather strong teeth that extend deeply into the aperture. Light brown on top, usually with two obscure transverse bands, and covered with white spots of different sizes. Grows to about 70 mm long. Indo-Pacific and northern Australia, extending well down into the temperate zone on both the east and west coasts.

LYNX COWRY *Cypraea lynx*

Oval shell with a rather flat base. Fawn or bluish grey on top, clouded with overlapping brown and mauve spots; large brown spots on the sides. Fawn or white base, orange between the strong teeth. Grows to about 50 mm long. Indo-Pacific and northern Australia.

TIGER COWRY *Cypraea tigris*

Probably the best known cowry. Heavy shell with coarse, short teeth. White or pale brown on top with large brown spots and a prominent pallial line. Grows to about 130 mm long. Coral reefs, often crawling in pools out in the open even during the day. Indo-Pacific and northern Australia.

Little Bear Stromb

Milk-spot Cowry

Lynx Cowry

Terebellum Conch

Tiger Cowry

ARABIC COWRY *Cypraea arabica*
Medium-sized with an oval shape and fine, brown teeth that cross the columella. Fawn, covered on top with a dense pattern of script-like brown marks, some enclosing clear spaces, and large spots on the sides. Prominent pallial line. Grows to 100 mm long. Indo-Pacific and northern Australia.

EGLANTINE COWRY *Cypraea eglantina*
Very like the Arabic Cowry but more slender and characterised by a distinct brown blotch at the spire. Grows to 100 mm long. Indo-Pacific and northern Australia.

CARNELIAN COWRY *Cypraea carneola*
Rather solid with fine teeth that extend well into the aperture. Orange-brown on top with four darker transverse bands; fawn sides with a smoky granular pattern. A defining feature is the deep purple or lilac teeth. May grow to 50 mm long. Indo-Pacific and northern Australia.

QUEEN ISABELL'S COWRY *Cypraea isabella*
Solid and nearly cylindrical with a narrow aperture and fine teeth. Fawn or grey on top with black axial lines and distinctive orange spots at the ends. Adult shells can be 20–40 mm long. Indo-Pacific and northern Australia.

SIEVE COWRY *Cypraea cribraria*
A very distinctive species of variable size. Specimens about 35 mm long would be average. Oval with short teeth. White sides and base but the top is dark red to brown with prominent white spots. Shape and the details of colouring are very variable and many forms have been named. Indo-Pacific and northern Australia.

HONEY COWRY *Cypraea helvola*
Small and solid with a row of pits along the callused margins. Coarse, short teeth extend over the columella. Reticulate pattern of darker lines and orange-brown spots on top; orange-brown sides and base. A feature is the presence of lilac spots at the ends. Grows to 35 mm long. Indo-Pacific; very common in the intertidal zone across northern Australia.

CAURICA COWRY *Cypraea caurica*
Subcylindrical with unusually big teeth. Greenish fawn with a large central brown blotch on top, plus three faint brown transverse bands. Grows to about 60 mm long. Indo-Pacific and northern Australia.

GREENISH COWRY *Cypraea subviridis*
Tumid and pyriform with short but moderately strong teeth. Greenish fawn on top, flecked with brown spots and crossed by a wide, irregular brown band; white, mauve or rose base. Grows to about 40 mm long. Indo-Pacific and northern Australia.

Arabic Cowry

Eglantine Cowry

Carnelian Cowry

Queen Isabell's Cowry

Sieve Cowry

Honey Cowry

Caurica Cowry

Greenish Cowry

CYLINDRICAL COWRY *Cypraea cylindrica*

As the name indicates, this cowry is cylindrical with fine teeth that extend deep into the aperture. Bluish white on top with fine brown freckles. There is usually a prominent brown dorsal blotch and a pair of brown spots at each end. Grows to 50 mm long. Indo-Pacific and northern Australia.

EROSE COWRY *Cypraea erosa*

Depressed cowry with thickened margins and very strong teeth that almost cross the base. Fawn or pale olive green on top with scattered pale brown spots and white dots. There are usually prominent brown blotches on the sides. Grows to about 50 mm long. Rocky and coral reefs. Indo-Pacific and northern Australia.

MILIARIS COWRY *Cypraea miliaris*

Tumid and pyriform with moderately strong teeth. Fawn or olive green on top with many tiny white spots and a prominent pallial line. Grows to about 50 mm long. Seems to prefer the more turbid waters of inshore areas. Indo-Pacific and northern Australia.

MONEY COWRY *Cypraea moneta*

Small, solid, somewhat depressed shell with irregularly nodulose sides and short, strong teeth. Yellow, orange or greenish grey on top with three faint darker bands and sometimes a faint encircling line. Grows to about 30 mm long. It was once used as a form of currency in some Indian and Pacific Ocean countries. Very common high in the intertidal zone of rocky shores. Widely distributed in the Indo-Pacific and northern Australia.

RINGED COWRY *Cypraea annulus*

Oval-shaped with short, strong teeth. Bluish white on top with a conspicuous yellow or orange encircling ring; cream, white, yellow or grey sides and base. Grows to 30 mm long. Abundant in the intertidal zone of coral reefs and rocky shores. Indo-Pacific and northern Australia.

WANDERING COWRY *Cypraea errones*

Characterised by its subpyriform shape and short, weak teeth. Greenish on top with three faint blue bands, brown freckles and a large central brown blotch; creamy white or yellow base and sides. Grows to 40 mm long. A common species that lives unusually high in the intertidal zone of rocky shores and coral reefs. Indo-Pacific and northern Australia.

SWALLOW COWRY *Cypraea hirundo*

Small and oval-shaped with fine teeth that extend almost all the way across the white base. Beautiful light blue on top with narrow pale blue bands, a brown dorsal blotch, scattered tiny brown spots and conspicuous pairs of brown spots at each end. Grows to about 20 mm long. Indo-Pacific and northern Australia.

Cylindrical Cowry

Erose Cowry

Miliaris Cowry

Money Cowry

Ringed Cowry

Wandering Cowry

Swallow Cowry

GRAPE COWRY *Cypraea staphylaea*

This small cowry and its relatives are unusual in having a nodulose top. This species has strong teeth that spread right across the base. Fawn or grey on top with white nodules; orange base and teeth; dark orange spots at the ends. In life the black mantle has many short but stout branches or filaments. Grows to 30 mm long. Indo-Pacific and northern Australia.

LIMACINA COWRY *Cypraea limacina*

More elongate than its relative, the Grape Cowry, the top is smooth or weakly nodulose and the teeth spread only part way across the base. Orange-brown or bluish grey on top with white spots; white base and teeth; orange ends. In life the red mantle has a profusion of long fleshy filaments. Grows to 35 mm long. Indo-Pacific and northern Australia.

PEPPERED COWRY *Cypraea (Notocypraea) piperita*

The notocypraeas form an endemic group in southern Australia. They live under stones on rocky shores. Their shells are cast up in hundreds on the shore after storms and, with their delicate colours, they are favourite pickings of beachcombers. This species is probably the most common. It is oval with a cream or pale tan top, crossed by four interrupted bands; usually profusely spotted with brown. Cream or fawn base and heavily spotted sides. Grows to about 25 mm long. From the south coast of New South Wales across southern Australia to the central west coast.

BROWN COWRY *Cypraea (Notocypraea) angustata*

Largest of the Notocypraea group; grows to 30 mm long. Oval to subpyriform. Uniform brown, tan or grey on top with an off-white base and heavily spotted sides. At the ends there are prominent spots on the sides of the canals. South-eastern Australia from southern New South Wales to St Vincent Gulf in South Australia.

COMPTON'S COWRY *Cypraea (Notocypraea) comptonii*

Slender with short, fine teeth. Plain orange-brown top with two faint transverse bands near the centre and sometimes faint bands at the ends. Sparsely spotted sides with brown blotches at the ends. Grows to 25 mm long. Southern Australia from southern New South Wales to Cape Leeuwin in Western Australia.

FLEA-BITTEN COWRY *Notocypraea pulicaria*

Small, elongate and subcylindrical with very fine teeth. Pink, rose or pale orange with large spots that sometimes merge to form transverse bands. Grows to about 20 mm long. Endemic to the south-western corner of Australia.

SPECKLED COWRY *Cypraea (Notocypraea) declivis*

Oval and humped with a milky white, fawn or sepia top profusely spotted with brown; sides strongly spotted. Adults are about 25 mm long. Endemic to the south-east, mainly in the Bass Strait.

Grape Cowry

Limacina Cowry

Peppered Cowry

Compton's Cowry

Brown Cowry

Flea-bitten Cowry

Speckled Cowry

EGG AND SPINDLE COWRIES Family OVULIDAE

Egg and spindle cowries are relatives of the true cowries but lack the strong apertural teeth of that group. The shells range from globular to extremely long and slender with long, drawn-out anterior and posterior canals. All the members of this family live in association with soft corals, anemones, gorgonians or hydrozoans, feeding on their polyps. The mantles of these elegant snails very often mimic both the colour and the surface texture of the host. Most species live in the tropics but there are a few in the temperate waters of southern coasts.

EGG COWRY *Ovula ovum*
The largest of the family; grows to about 100 mm long. Inflated and egg-shaped. Polished, milky white exterior; orange-brown interior. Contrasting with the shell, the animal is black. Lives and feeds on the large, fleshy soft coral *Sarcophyton*. Widely distributed in the Indo-Pacific and across northern Australia.

SPINDLE COWRY *Volva volva*
Thin and elegant with a smooth, inflated body whorl and extremely long, curved anterior and posterior canals. Cream or flesh-pink exterior; light brown interior. Grows to 100 mm long from tip to tip. Large sea-whips in the subtidal zone. Indo-Pacific and northern Australia.

UMBILICAL EGG COWRY *Calpurnus verrucosus*
Solid, oval shell with a flat base and conspicuous button-like knobs above the short anterior and posterior canals. Minutely grooved and white on top with pink ends. Grows to about 35 mm long. Lives on the fleshy soft coral *Lobophytum*. Indo-Pacific and Queensland.

TOKIO'S SHUTTLE *Phenacovolva tokioi*
The shuttles are slender, fusiform shells with long, pointed canals. This species has a pink, smooth or minutely grooved exterior, usually with a central white transverse band. Grows to about 40 mm long. Lives on seawhips. Central Indo-Pacific and northern Australia.

PYRIFORM EGG COWRY *Margovula pyriformis*
A moderately solid little shell that has a pyriform shape and a minutely spirally striate surface. Outer lip toothed, inner lip smooth and callused only at the ends. Fawn to pale brown with an obscure, darker double band across the back; ends sometimes orange. Animal black. The shell may be up to 22 mm long. This species is reported to live on anemones in the shallow subtidal zone. North-eastern Australia.

Egg Cowry

Two Egg Cowry specimens, their black mantles spread over their shells, feeding on a soft coral

Spindle Cowry

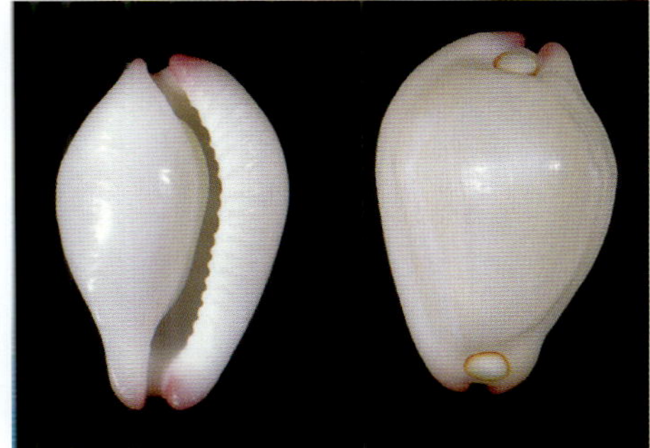

Umbilical Egg Cowry

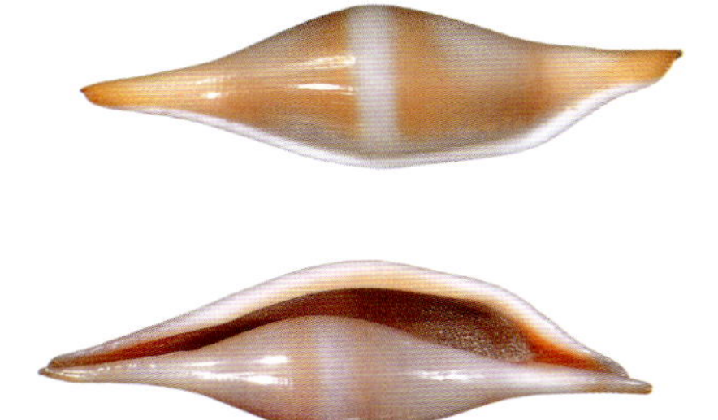

Tokio's Shuttle

Pyriform Egg Cowry

BEAN COWRIES Family TRIVIIDAE

These little shells resemble cowries but they have transverse ribs across the back. There are at least 10 Australian species. Two common species have been chosen to illustrate the group.

SULCATE BEAN COWRY *Trivia oryza*
White and oval with slightly beaked ends. A prominent groove on the back interrupts the transverse ribs. Grows to 10 mm long. Under stones in the intertidal zone. Indo-Pacific; northern Australia.

AUSTRALIAN BEAN COWRY *Trivia merces*
Small, oval, strongly ridged shell with blunt ends. Prominent rose-coloured blotches on the back. Grows to 13 mm long. Very common among debris on beaches across southern Australia.

MOON SNAILS Family Naticidae

These carnivorous snails have large fleshy bodies. The meandering trails they make as they wander about hunting bivalves are a feature of intertidal sand flats. They drill tiny holes through the shells of their prey. Females lay egg masses of stiffened jelly in which many tiny eggs are embedded, forming the sandy egg collars that are so common on sand flats. The operculum may be thin and horny (*Polinices* and *Sinum*) or thickened with calcium carbonate (*Natica*).

CONICAL MOON SNAIL *Polinices conicus*
High-spired, inflated shells with a narrow umbilicus. Cream, fawn or grey-blue with a brown or orange subsutural band and columella; chocolate brown callus and umbilicus. Grows to about 50 mm high. Common on sand flats across southern Australia.

SORDID MOON SNAIL *Polinices sordidus*
Like the previous species, and about the same size but more globular. Tan or blue-grey with an orange subsutural band. Orange or brown columellar callus and umbilicus; chocolate-brown aperture. Common on muddy sand flats. Eastern and south-eastern Australia.

INCE'S MOON SNAIL *Polinices incei*
Flattened snail with a low spire and a button-like callus almost filling the umbilicus. White, pale yellow, grey, brown or purple on top; usually white on the base with a brown callus. Grows to 30 mm wide. Muddy flats. Eastern and south-eastern Australia.

BLADDER MOON SNAIL *Polinices didyma*
Distinguished by its large, globular shell with a low spire. Deep umbilicus which is almost closed by a deeply grooved, thick, tongue-like callus. Blue-grey or fawn exterior with a narrow orange subsutural band; brown interior, callus and columella. Grows to 70 mm high. Common on muddy flats. Central Indo-Pacific and northern Australia.

Sulcate Bean Cowry

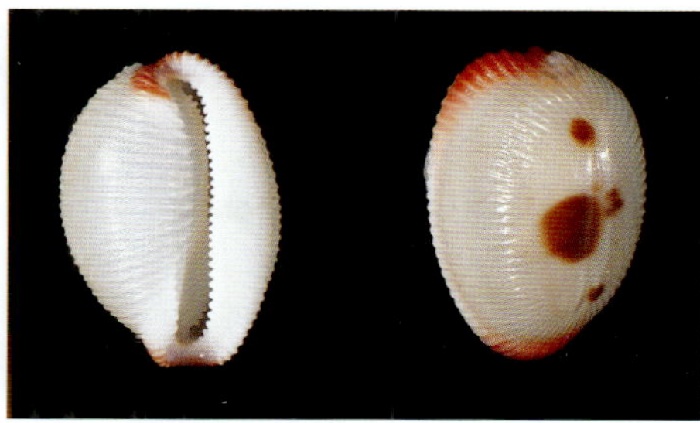

Australian Bean Cowry

Sordid Moon Snail

Conical Moon Snail

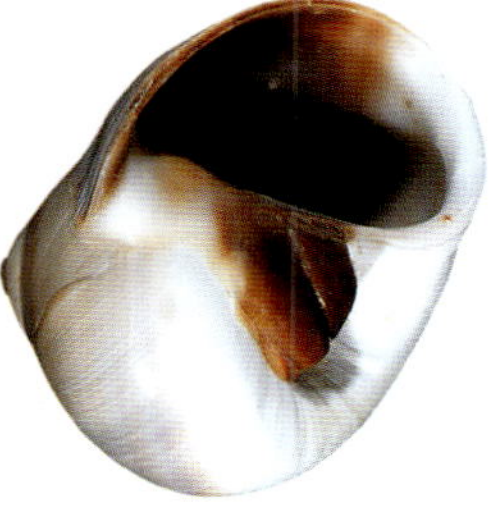

Ince's Moon Snail

Bladder Moon Snail

UMBILICATE MOON SNAIL *Polinices powisiana*
A large shell, characterised by a very wide umbilicus and strong funicle. Glossy and white with a broad orange-brown band. Grows to 60 mm high. Indo-Pacific and northern Australia.

GUALTIERI'S NATICA *Natica gualtieriana*
Globular with a low spire and short axial folds below the suture. Greenish or grey with two spiral bands of short brown lines. Smooth operculum with a single groove around the margin. Very common and widely distributed on intertidal sand flats. Grows to about 30 mm high. Indo-Pacific and across northern Australia.

BLACK-MOUTH MOON SNAIL *Polinices melanostomus*
Thin with a low spire and oblique whorls. Glossy off-white or fawn exterior with darker spiral bands; dark red brown interior and columellar callus. Grows to about 40 mm high. Coral reefs. Indo-Pacific and northern Australia.

LINED NATICA *Natica lineata*
Handsome, globular shell with a thick funicle and a double-grooved operculum. Cream outer surface with thin, red or orange axial lines; operculum has a prominent brown blotch at the apex. Grows to 40 mm high. Muddy areas. Indo-Pacific and the far north of Australia from the Kimberley in Western Australia to the Torres Strait.

SOLID NATICA *Natica fasciata*
Globular with a low spire and partly closed umbilicus. Operculum has a single spiral groove and serrated edge. Fawn to dark brown; when light-coloured there are wide bands of brown. Dark brown callus, white operculum. Grows to 30 mm high. Intertidal sand flats. Indo-Pacific and northern Australia.

CALF NATICA *Natica vitellus*
Very striking globular shell with a low spire, flaring anterior end, partly closed umbilicus and small funicle. Operculum has two spiral grooves. Orange or yellow with one or two irregular white bands. Grows to 40 mm high. Sand in the intertidal and shallow subtidal zones. Indo-Pacific and northern Australia.

JAVANESE EAR MOON *Sinum javanicum*
Ear moons are thin-shelled, depressed and ear-shaped like abalones. The animal is very big and slug-like. This species has a cream or white shell with a fine sculpture of spiral cords on the top. Grows to 35 mm wide. Sand in the intertidal and shallow subtidal zones. Central Indo-Pacific and northern Australia.

Umbilicate Moon Snail

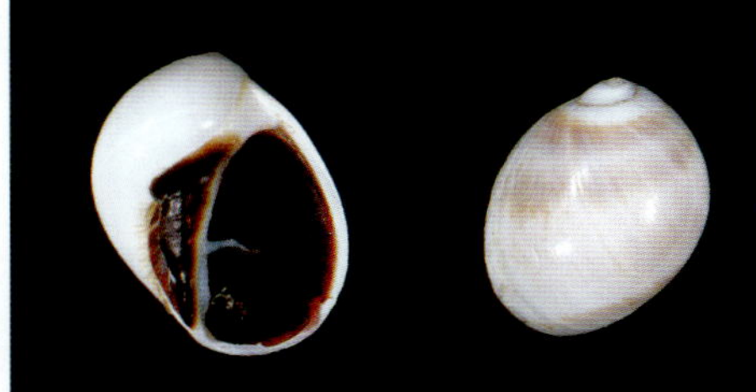

Black-mouth Moon Snail

Lined Natica

Calf Natica

Gualtieri's Natica

Solid Natica

Javanese Ear Moon

HELMET AND BONNET SHELLS Family CASSIDAE

Some of these heavy shells are said to resemble Roman helmets; others are supposed to resemble bonnets, hence the names. These sand-dwelling carnivores feed on echinoderms. The shells are usually varicose with a thickened outer lip and a callus shield on the columellar side. The operculum is oval and horny. When the animal is out hunting the large flat foot is expanded and the siphon is held erect as it seeks out the 'smell' of the prey.

HORNED HELMET *Cassis cornuta*
This huge shell grows to 350 mm long with heavy horn-like tubercles on the shoulder. Usually worn and grey on top with a flat, polished, yellow-orange base. Sand in coral reef lagoons. This species is protected in Australia. Indo-Pacific and northern Australia.

FIMBRIATE HELMET *Cassis fimbriata*
Usually sculptured with axial ribs and two or three spiral rows of shoulder tubercles. Shield lirate with a smooth or weakly toothed outer lip. Shiny cream exterior with axial brown patches and interrupted brown spiral lines. Grows to 120 mm long. Sand mainly in the subtidal zone. Southern Australia from Bass Strait to Geraldton in Western Australia.

LIPPED BONNET *Semicassis labiata*
Smooth and polished with rounded or weakly nodulose shoulders, toothed outer lip and weakly lirate columella. Cream exterior with purple-brown blotches and brown anterior canal. Grows to 80 mm long. Subtidal. Circum-Southern Ocean, eastern and southern Australia.

BANDED BONNET *Phalium bandatum*
Oval shell similar to the Checkerboard Bonnet but with a taller spire. Creamy exterior with five bands of yellow-brown blotches; six orange-tan blotches on the lip; light orange interior and shield. Grows to 100 mm long. Subtidal but often cast ashore after storms. Central Indo-Pacific and northern Australia, extending down east and west coasts into temperate latitudes.

CHECKERBOARD BONNET *Phalium areola*
Oval but tending to be quadrate and squared at the shoulder. Tall, pointed spire, prominent varices and a wide outer lip bearing teeth along its inner edge. Five rows of well-defined, squarish dark brown spots. White, lirate columellar shield; deep interior brown. Grows to about 120 mm long. Subtidal. Central Indo-Pacific and northern Australia.

HEAVY BONNET *Casmaria ponderosa*
Variable. Most shells are heavy with a high spire, smooth or nodulose shoulders and a thick lip bearing prickle-like teeth and dark brown spots. Glossy, white, grey, cream or pale tan exterior with spiral bands of prominent brown spots below the suture, another behind the anterior canal, and a large brown blotch at the base of the siphonal notch. Grows to 70 mm long. Sand. Eastern Pacific, Indo-Pacific and northern Australia.

Horned Helmet

Fimbriate Helmet

Lipped Bonnet

Banded Bonnet

Checkerboard Bonnet

Heavy Bonnet

VIBEX BONNET *Casmaria erinacea*

A confusing species because some specimens (like the one illustrated) have smooth shoulders, while others have heavily nodulose shoulders. The cream or fawn exterior usually has axial rows of tiny brown spots and sometimes axial lines or bands. Spiny outer lip is marked by large brown spots. Grows to about 70 mm long. Sand in the intertidal and subtidal zones. Indo-Pacific and northern Australia.

ANGAS' BONNET *Semicassis angasi*

Inflated shell that is pale pink, white or cream, apart from a few indistinct pale tan patches at the suture. Toothed outer lip. Grows to 60 mm long. Sand cays associated with coral reefs. A distinctly Australian shell named after an early Australian conchologist. Northern Australia.

PEAR BONNET *Semicassis pyrum*

Rather globular but with a moderately high spire and nodulose shoulders. The narrow outer lip lacks teeth and the columellar shield is smooth. Cream or fawn with spiral bands of diffuse brown spots or stains; anterior canal tipped with brown. Grows to 70 mm long. Southern Australia from New South Wales to Fremantle; a form of the species also occurs in New Zealand.

TRITONS AND TRUMPETS Family Ranellidae

This family of carnivorous snails has a variety of diets, including bivalves, gastropods, sea squirts and echinoderms. They have thick shells, strong varices and a long anterior canal that forms a spout. The columella is usually glazed, the glaze often forming an adherent or raised shield. The larvae of most ranellids hatch as free-swimming veligers and there is a long planktotrophic larval stage. As a result many species have a very wide geographic range.

AUSTRALIAN RED TRITON *Charonia lampas*

Like the Trumpet Triton but much smaller with a thicker shell. Yellow-brown exterior with red-brown blotches, dark spots and white nodules. White interior; brown columella with white lirae. Grows to 150 mm. Widespread in middle latitudes around the world. The Australian representatives have several distinctive geographic forms. Rocky shores in the intertidal and subtidal zones. Southern Australia.

TRUMPET TRITON *Charonia tritonis*

This remarkable shell grows to 450 mm long. It has convex, spirally ribbed whorls which are fawn or light brown with darker purple-brown crescent-shaped markings. Orange-yellow interior; columella dark brown between white or yellow lirae. Sharply toothed outer lip. Islanders break a hole in the tall, slender spire and use this shell as a trumpet. This triton eats starfish, including the infamous Crown-of-Thorns. Coral and rocky reefs from the intertidal zone to the shallow subtidal zone. Protected in Australian waters. Indo-Pacific and northern Australia.

Vibex Bonnet

Angas' Bonnet

Pear Bonnet

Australian Red Triton

Trumpet Triton

AUSTRAL TRITON *Ranella australasia*

Sculptured with spiral threads and ribs. Beneath a furry periostracum the exterior is usually red-brown or cream with dark brown and white bands on the strong varices. White interior. Grows to 90 mm long. Rocky shores. Southern Australia.

SPENGLER'S TRITON *Cabestana spengleri*

Medium-sized and notable for its strong spiral ribs and toothed lip. Smooth, reflected columella. Beneath a thick periostracum the exterior is yellow-brown, usually darker between the ribs; white interior and columella. Grows to 150 mm long. Among sea squirts on rocky shores. Eastern and south-eastern Australia.

NEOPOLITAN TRITON *Cymatium parthenopeum*

Broader than others of the family with convex whorls sculptured with nodulose spiral ribs and axial striae. Toothed lip and lirate columella. Beneath a thick, hairy periostracum the exterior is light yellow-brown. Grows to 140 mm long. Widely distributed on rocky shores. Cosmopolitan in middle latitudes; southern Australia.

HAIRY TRITON *Cymatium pileare*

Slender and varicose with flat, knobby spiral ribs and toothed lip. Columella heavily lirate. Beneath a hairy periostracum the exterior is cream, blue-grey or fawn with brown axial streaks and spiral bands. Orange or red interior and columella. Grows to about 100 mm long. Common under stones on rocky shores and coral reefs. Indo-Pacific and northern Australia.

BRISTLED TRITON *Cymatium vespaceum*

Tall-spired with a long, pointed anterior canal and spirally ribbed, heavily nodulose whorls. Cream or brown with a dark brown blotch at the end of the anterior canal. In life these tritons have a bristled periostracum. Grows to about 60 mm long. Common on intertidal rocky flats where they feed on sea squirts. Indo-Pacific and northern Australia, extending far down the east and west coasts into temperate latitudes.

PARKINSON'S SASSIA *Sassia parkinsonia*

This little shell is the most common of four Australian species of a genus that is widespread throughout the world. Solid and chunky with a tall spire and heavily variced and nodular whorls sculptured with minute spiral threads. Yellow-brown exterior, white interior. Grows to 45 mm long. Common in the intertidal and shallow subtidal zones. South-eastern Australia.

Austral Triton

Spengler's Triton

Neopolitan Triton

Hairy Triton

Bristled Triton

Parkinson's Sassia

TUN SHELLS Family TONNIDAE

These large carnivorous snails feed on sea cucumbers. They live in sand and are capable of rapid crawling by means of a large flat foot. Most shells are inflated and sculptured with spiral ribs. There is a thin periostracum. Egg masses consist of wide gelatinous ribbons containing small, transparent eggs. The veligers hatch as planktotrophic larvae and there is a very long development time. Some species of the family are very widely distributed by ocean currents.

GRINNING TUN *Malea pomum*

These shells take their common name from their wide mouth and toothed lip. They are heavier than most in the family and sculptured with rounded spiral ribs and plaited columella. Fawn or cream exterior with large white spots on the ribs. Yellow interior; white or cream columella and lip. Grows to about 60 mm long. Sand on coral reefs. Indo-Pacific and northern Australia.

CHINA TUN *Tonna chinensis*

One of the smaller tuns; grows to about 60 mm long. Globular and moderately thin with a simple lip and reflected columellar callus. Sculptured with rounded spiral ribs. Yellow-brown or cream exterior with white dashes and brown spots. Sandy pools on rocky or coral reefs. Indo-Pacific and northern Australia.

VARIEGATED TUN *Tonna variegata*

The largest of the Australian tuns; grows to about 200 mm long. Globular and thin-shelled with a low spire, thin, simple lip and about 17 broad rounded spiral ribs on the body whorl. Yellow cream or light brown exterior; some ribs are white or pale cream with brown spots. Southern Australia. Also found in New Zealand.

PARTRIDGE TUN *Tonna perdix*

Globular and thin-shelled with a relatively high spire and thin, simple lip. Light brown exterior is sculptured with low spiral ribs bearing crescent-shaped white markings. Yellow-brown interior. Grows to about 140 mm long. Widespread Indo-Pacific species also found across northern Australia.

TESSELATE TUN *Tonna tessellata*

Thin and almost spherical, sculptured with strong spiral ribs. Unusually for the genus, in fully adult shells the lip is crenulate along its outer edge and toothed within. Exterior is usually fawn with white ribs bearing orange-brown spots. While the edge of the lip is white, the deep interior is yellow. Grows to about 100 mm long. Indo-Pacific and northern Australia.

BANDED TUN *Tonna sulcosa*

Globular and sculptured with flat spiral cords; outer lip weakly crenulate . Notable for the dramatic spiral brown bands around the otherwise white shell. Grows quite large — to 120 mm long. Indo-Pacific and Queensland.

Grinning Tun

China Tun

Variegated Tun

Partridge Tun

Tesselate Tun

Banded Tun

WENTLETRAPS AND VIOLET SEA SNAILS
Families EPITONIIDAE and JANTHINIDAE

Although their shells are not similar, these two families share many features, including their habit of preying on coelenterates; they are classified together. Wentletrap shells are usually tall with many convex, axially ribbed whorls. Most species live in sand associated with anemones. They are white or fawn with white ribs; their bodies sometimes match the colour of the host. The fragile violet sea snails are globular and coloured deep violet. They have a very unusual natural history, spending their lives suspended by a mucous float at the surface of the sea. They are associated with the pelagic coelenterates of the genus *Velella*.

IMPERIAL WENTLETRAP *Epitonium imperialis*
Though stout with tumid whorls, this is a delicate, thin-shelled species with thin, crowded ribs and deep suture and umbilicus. Pale fawn to brown between the white ribs. Grows to about 40 mm long. Associated with sea anemones on sand flats. Indo-Pacific and northern Australia.

PERPLEXING WENTLETRAP *Epitonium perplexum*
Tumid, loosely coiled whorls and a deep suture; no umbilicus. Sharp-edged, white axial ribs — about 13 on the last whorl — with a thin spiral cord around the base. White or fawn with obscure reddish spiral bands. Grows to 40 mm long. Associated with sea anemones on sand flats. Indo-Pacific and northern Australia.

PRECIOUS WENTLETRAP *Epitonium scalare*
Stout with a deep suture and umbilicus. The separated whorls are connected by widely spaced axial ribs. White to pale grey or fawn with white ribs. Grows to about 50 mm long. This remarkable shell was regarded as a treasure by eighteenth-century collectors, who paid large prices for specimens — hence its common name. Associated with sea anemones on sand flats. Indo-Pacific and Queensland.

PALLAS' WENTLETRAP *Epitonium pallasi*
Stouter than most others with thin, widely spaced axial ribs. Fawn or pale brown. Grows to about 25 mm long. Indo-Pacific; recorded in northern Australia from Darwin to Port Douglas.

AUSTRAL WENTLETRAP *Opalia australis*
Solid, white, tall-spired, slender, elongate shell with tightly fused whorls. No umbilicus; axial ribs end at a spiral rib around the base of the last whorl. Grows to about 40 mm long. Very common in beach drift along the southern coast. Southern Australia from Sydney to Fremantle.

GRANOSE WENTLETRAP *Opalia granosa*
Tall and slender like the previous species and about the same size but lacking axial ribs; short folds beneath the suture. White. Very common on beaches. Southern Australia from Bass Strait to Fremantle.

Imperial Wentletrap

Precious Wentletrap

Austral Wentletrap

Perplexing Wentletrap

Pallas' Wentletrap

Granose Wentletrap

COMMON VIOLET SEA SNAIL *Janthina janthina*

The largest and most common species of its group. Fragile, turbinate shell with a shallow indentation in the lip edge. Columella and outer lip meet at a sharp corner. Violet, paler near the suture. Grows to about 30 mm wide. Very common on southern beaches after winter storms, often still attached to its jelly bubble float. Cosmopolitan; southern Australia.

MUREX AND ROCK SHELLS Family MURICIDAE

This is a very big family that is divided into several distinct subfamilies. True murex shells (Muricinae) are most notable for their prominent varices and spiny or leafy sculpture. The varices probably strengthen the shell while the spines and fronds serve for protection. Rock shells (Thaidinae) and oyster drills (Ergalataxinae) do not have such prominent sculpture. Muricids are carnivorous, feeding on a range of invertebrate prey such as other molluscs, barnacles and worms. Many mollusc-eating species attack their prey by drilling holes in their shells by means of a rotating radula and shell-softening secretions. One subfamily, the purple-mouthed coral shells (Coralliophilinae) live associated with corals and feed on their polyps.

SPINY MUREX *Murex acanthostephes*

Species of the genus *Murex* have very long anterior canals and three spiky varices. There are several Australian species. This one grows to about 110 mm long. Each varix has three curved primary spines with smaller secondaries in between, plus additional spines on the canal. Cream or fawn with white flecks, spine tips dark grey. Sand in the intertidal and subtidal zones. An endemic northern Australian species ranging from North West Cape to the Torres Strait.

SHORT-SPINED MUREX *Murex brevispina*

Similar in form to the previous species but with strongly shouldered whorls and fewer, shorter curved spines. Yellow-brown or bluish grey exterior, often with three diffuse brown bands; brown inside the aperture but white around the mouth. Grows to about 80 mm long. Sand in the intertidal and subtidal zones. Indo-Pacific and northern Australia.

THIN-SPINED MUREX *Murex tenuirostrum*

Delicate with a long, thin anterior canal and long, thin spines: the spine on the dorsal varix may be very long and curved over the tall spire. Delicate ivory to golden brown, sometimes with red-brown blotches behind the notches on the lip. Grows to 140 mm long. Sand in the intertidal and subtidal zones. Central Indo-Pacific and along the east coast of Australia as far south as Tin Can Bay in Queensland.

Common Violet Sea Snail

Spiny Murex

Short-spined Murex

Thin-spined Murex

TRIPLE MUREX *Pterynotus triformis*

This species and its relatives have tall spires, long anterior canals and three high, scaly lamellate varices on each whorl, each extending to the tip of the canal. This species has a white, yellow, pink or brown exterior and a white interior. Grows to 60 mm long. Sand among rocks and seagrass in the intertidal and subtidal zones. Southern Australia.

MONODON MUREX *Chicoreus cornucervi*

High, conical spire, curved canal. Tumid body whorl roughly sculptured with spiral cords and three varices bearing long curved frondose spines. Brown to fawn exterior; pink around the rim of the mouth. Grows to 110 mm long from tip to tip. On rocks in the intertidal zone. North West Cape to the Torres Strait.

TRIVIAL MUREX *Chicoreus trivialis*

One of the smaller members of the genus; grows to about 70 mm long. Tall spire, tapering canal and three varices per whorl bearing low leafy fronds. Brown with a pale band below the suture and pink around the aperture. Under stones in the shallows. Central Indo-Pacific; restricted Australian distribution in the Kimberley, Western Australia.

CURLY MUREX *Chicoreus microphyllus*

Fusiform and tall-spired. Each whorl has three varices that bear short, leafy spines with minor spines between. Brown exterior; white interior with yellow or orange columella and lip. Grows to about 110 mm long. On rocks in the subtidal zone. Indo-Pacific and northern Australia.

PURPLE-MOUTHED DRUPE *Drupa morum*

Very thick and nodulose with a low spire and flat base. The lip bears strong marginal spines and strong inner teeth, some of them fused together. The columella has four strong, thick folds. White exterior with dark brown nodules and spines; dark violet interior. Grows to 40 mm long. Rocky shores in the lower intertidal zone. Indo-Pacific and northern Australia.

PRICKLY DRUPE *Drupa ricinus*

Like the Purple-mouthed Drupe but smaller and not so heavy. White exterior with black-tipped spines. The shiny interior is also white but usually has yellow blotches on the lip and columella. Grows to 30 mm long. Lower intertidal zone of rocky shores and coral reefs. Widely distributed in the Indo-Pacific and northern Australia.

GRANULATED DRUPE *Morula granulata*

Solid, biconical shell with five spiral rows of heavy tubercles around the body whorl. Narrow mouth with four teeth on the thick outer lip and two or more weak folds on the callused columella. White or grey on the outside with dark brown nodules and white teeth. Grows to 30 mm long. Very common in the upper intertidal zone of rocky shores, usually associated with rock oysters. Indo-Pacific and northern Australia.

Triple Murex

Monodon Murex

Trivial Murex

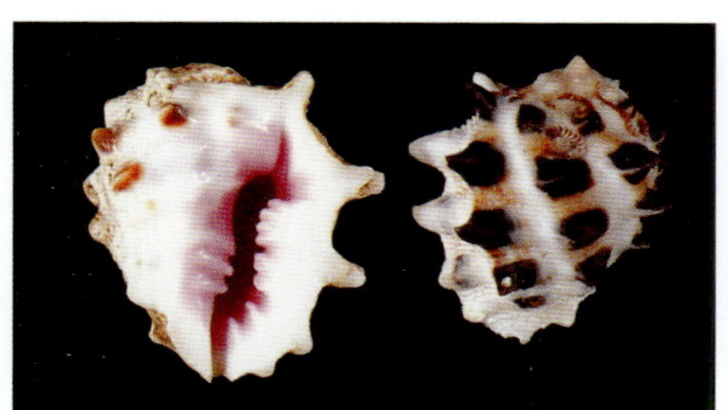

Purple-mouthed Drupe

Curly Murex

Granulated Drupe

Prickly Drupe

GRAPE DRUPE *Morula uva*
Small, solid biconical shell with prominent teeth on the lip and spiral rows of pointed nodules on the body whorl. White exterior with black nodules and a violet aperture. Grows to 30 mm long. Very common under rocks and corals in the intertidal and shallow subtidal zones. Indo-Pacific and northern Australia.

KNOBBY DRUPE *Drupella cornus*
Dense aggregations of this carnivorous snail sometimes do great damage to living corals — they eat the living polyps. Heavy, thick, nodulose shell, unremarkable except for its coral-eating habits. Cream exterior (though usually heavily encrusted), yellow interior. Grows to about 40 mm long. Widely distributed and abundant on coral reefs throughout the Indo-Pacific and in northern Australia from the Abrolhos Islands in the west to southern Queensland.

TRITON WHELK *Agnewia tritoniformis*
Small, fusiform shell with a very tall spire. Sculptured with weakly nodulose axial ribs and spiral threads. Usually has a cream exterior; white within the aperture. Grows to about 25 mm long. Very common on rocky shores in south-eastern Australia and New Zealand.

WESTERN DRILL *Cronia avellana*
Oval with a wide mouth and simple lip. White, often with two dark brown spiral bands and additional blotches. Grows to about 35 mm long. Very common predatory snail on rocky intertidal shores of Western Australia from the south coast to the Kimberley.

INDIAN NASSA *Nassa francolina*
Oval and tumid with a tall spire, wide mouth and simple lip. Sculpture of fine, spiral threads. Blotchy red-brown exterior with a paler central band; cream interior. Grows to about 70 mm long. Under stones in the intertidal and shallow subtidal zones of rocky and coral reefs. This is the Indian Ocean form of the genus, common on the Western Australian coast north of the Abrolhos. In Queensland it is replaced by the similar but nodulose Pacific species *N. serta*.

SOUTHERN ROCK SHELL *Thais orbita*
Heavy shell, oval in outline with extremely variable sculpture. The typical east coast '*orbita*' form is strongly ribbed, the '*textilosa*' form from the south-east has weak spiral sculpture , while the '*aegrota*' form (illustrated) from Western Australia is nodulose. Creamy exterior; white interior with a yellow tinge around the mouth. Grows to about 80 mm long. Rocky shores. Southern Australia. Another form of this species lives in New Zealand.

Grape Drupe

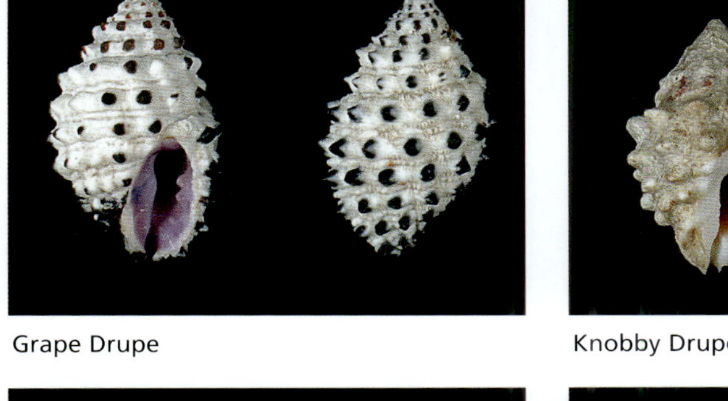

Knobby Drupe

Triton Whelk

Western Drill

Indian Nassa

Southern Rock Shell

ACULEATE ROCK SHELL *Thais aculeata*

Heavy and biconical with four spiral rows of stout conical nodules. The lip bears inner teeth extending into the aperture as lirae. Black exterior with white patches and streaks. Deep inside the aperture is bluish white but the rim of the mouth is usually purple-brown. Grows to about 60 mm long. Common among rock oysters in the upper intertidal zone. Indo-Pacific and northern Australia.

PRICKLY ROCK SHELL *Thais echinata*

Similar to the species above, but not so heavy, with pointed nodules. Creamy grey to fawn exterior, glossy white interior. Grows to about 50 mm long. Rocky shores and coral reefs. Indo-Pacific and northern Australia.

GIANT CONCHS Family TURBINELLIDAE

The snails of this family are scavengers or predators on bivalves. Many of the species are deep-sea dwellers but some, like the Australian Giant Conch *Syrinx*, live in the shallows. Females lay tough, flattened capsules, held together by a connecting strand. The capsules are stacked up together and have overlapping edges like a concertina. Development is direct.

AUSTRALIAN GIANT CONCH *Syrinx aruanus*

This is the largest living gastropod; grows to 600 mm long. Strong, fusiform shell with a straight-sided or turreted spire and a long, straight anterior canal. The very tall, multispiral protoconch is usually lost in adults. Light yellow-brown exterior covered by a thick periostracum; yellow, glossy interior. Sand in the intertidal and shallow subtidal zones. Northern Australia.

WHELKS Family BUCCINIDAE

Collectively known as whelks, the Buccinidae family includes a variety of genera that have their own common names. Most are tall-spired and have a moderately developed anterior canal. They range from small to large in size. They are carnivorous snails, some preying on other molluscs while others are scavengers.

LINEATED COMINELLA *Cominella lineolata*

Elongate–oval shell with rounded shoulders and weak sculpture. Variable colour: exterior usually yellow-brown with brown spiral lines or spots; brown interior; white columella with a brown patch. Grows to about 25 mm long. Very common in the intertidal zone of rocky shores. Southern Australia from New South Wales to the south coast of Western Australia.

Aculeate Rock Shell

Prickly Rock Shell

Australian Giant Conch (juvenile with protoconch intact)

Lineated Cominella

Australian Giant Conch

RIBBED COMINELLA *Cominella eburnea*

Elongate with a tall spire, angular shoulders and axially ribbed whorls. Yellow-brown to grey exterior with irregular brown blotches; deep interior is white, the lip and columella brown. Grows to 40 mm long. Sandy and muddy flats of bays and estuaries. Southern Australia from New South Wales to Geraldton in Western Australia.

NODULOSE COMINELLA *Cominella acutinodosa*

Biconical with angular, heavily nodulose shoulders. Fawn to yellow, flecked with brown spots on the spiral cords; purple-brown interior. Grows to 30 mm long. Very common on intertidal muddy sand flats on the Pilbara and Kimberley coasts of Western Australia.

PACIFIC PHOS *Phos senticosus*

The tropical phos species are oval shells with tall spires and short, twisted anterior canals. This is the most common one. It has no varices but sharp axial and spiral ribs forming prickles at the intersections and there is a plaited columella. Cream or fawn exterior with brown bands; purple or mauve interior and columella. Grows to 35 mm long. Sand in the intertidal and shallow subtidal zones. Indo-Pacific and northern Australia.

ORANGE-MOUTHED GOBLET *Cantharus erythrostomus*

Stoutly oval with stepped shoulders and a relatively short anterior canal. Sculptured with broad axial folds crossed by spiral ribs and axial striae. Generally orange-brown with darker brown ribs. The edges of the mouth are orange. Grows to about 40 mm long. Common under stones on rocky shores and coral reefs. Indo-Pacific and northern Australia.

WAVY GOBLET *Cantharus undosus*

Very solid shell, like the species above, but lacking axial ribs; instead it has strong spiral ribs. The ground colour is white but the close ribs are black or dark purple and the lip edge and columella are orange. Grows to about 40 mm long. Under stones on rocky shores and coral reefs. Indo-Pacific and northern Australia.

AUSTRALIAN PERISTERNIA *Peristernia australiensis*

Stoutly biconical with strong axial folds and spiral ribs. White exterior with conspicuous brown blotches between the ribs; brown or mauve interior and columella. Grows to about 35 mm long. A common coral reef species but apparently confined to north-eastern Australia.

FLESHY PERISTERNIA *Peristernia incarnata*

Elongate and fusiform with prominent axial folds and spiral ribs in the interspaces. Orange-brown exterior, darker in the interspaces; pink or mauve interior. Grows to 30 mm long. Very common under stones on intertidal rocky and coral reefs. Indo-Pacific and northern Australia from Geraldton in Western Australia to northern New South Wales.

Ribbed Cominella

Nodulose Cominella

Pacific Phos

Orange-mouthed Goblet

Wavy Goblet

Australian Peristernia

Fleshy Feristernia

FLAME PISANIA *Pisania ignea*

Elongate and fusiform with rounded, smooth whorls. These shells often lose the top of the spire ('decollate'). Cream exterior with brown axial flames and a pale central band; purple interior. Grows to 35 mm long. Under stones on rocky and coral reefs. Widely distributed in the Indo-Pacific and across northern Australia.

MELON CONCH *Pugilina cochlidium*

The common name comes from the subfamily Melongeninae (family Melongenidae in some classifications) of which this is the sole Australian representative. Heavy shell with angular shoulders bearing pointed nodules, a wide canal and curved but simple columella. Uniformly orange-brown. Grows to 120 mm long. Intertidal muddy flats. Indian Ocean species found across the far north of Australia from the Kimberley to northern Queensland.

SPINDLES AND THEIR RELATIVES
Family FASCIOLARIIDAE

The family Fasciolariidae, whose shells are known as spindles and horse conchs, is characterised by a long anterior canal. The operculum is thin, clawed and horny. The animals of this group are red-pigmented. Like their relatives, the whelks, they are carnivorous snails, mostly preying on other molluscs.

SOUTHERN SPINDLE *Fusinus australis*

The spindles are very elongate and fusiform with tall spires and long, straight anterior canals. This species is sculptured with axial folds and narrow spiral cords. Brown exterior beneath a thick, brown periostracum; white interior. Grows to about 110 mm long. Often associated with seagrass beds. Common across southern Australia from Bass Strait to Fremantle.

DISTAFF SPINDLE *Fusinus colus*

Slender with a very long anterior canal. The shoulders of the whorls may be rounded or angular. White but usually with brown patches between nodules on the spire and at the tip of the canal. Grows to 140 mm long. Sand-dweller whose shells are commonly found on beaches but not often seen alive in the shallows. Indo-Pacific and northern Australia.

AUSTRALIAN HORSE CONCH *Pleuroploca australasia*

This is an extremely variable species and many forms have been separately named. The common shallow water form is broadly fusiform with a turreted spire, nodulose shoulders and a moderately long anterior canal. Light brown exterior beneath a thin, olive-brown periostracum. Grows to about 150 mm long. Among algal-covered rocks. Southern Australia.

Flame Pisania

Southern Spindle

Australian Horse Conch

Melon Conch

Distaff Spindle

FILAMENTOUS HORSE CONCH *Pleuroploca filamentosa*

Elegant shell with a tall spire and a long canal. Finely sculptured with thin spiral ribs. Dark brown with cream patches on the shoulders — some ribs may also be cream; orange or cream interior. May grow to 110 mm long. Coral reef species that hides under stones during the day. Indo-Pacific and northern Australia.

RECURVED LATIRUS *Latirus recurvirostris*

Latirus shells are generally fusiform with a long anterior canal and a thickly folded columella. This one has angular shoulders and prominent axial folds with transverse, sharp-edged nodules where the spiral ribs cross. The thick canal turns a little to the right. Orange-brown both inside and out but with dark brown blotches between the nodules. Lives among rocks in the subtidal zone. Grows to about 80 mm long. Indo-Pacific and northern Australia.

TOWER LATIRUS *Latirus turritus*

This common species is smaller than most in the genus, most specimens being about 40 mm long. It has an unusually high spire, short canal and sculpture of axial folds and spiral ribs. Red-brown exterior encircled by black spiral lines on the ribs; orange interior. Common under stones on intertidal rocky and coral reefs. Indo-Pacific and northern Australia.

DOG WHELKS Family NASSARIIDAE

Dog whelks are relatively small snails with oval shells. They are carnivores or scavengers. The animals have a large foot with a pointed tail end. The siphon is long and flexible but the shell has no anterior canal, only a deep u-shaped notch at the anterior end through which the siphon may be extended when the animal is hunting. There is also a deep anal notch at the top end of the aperture. These active snails live on sand or mud and can crawl quite rapidly but they slide over the surface and do not make trails like those of the moon snails. Some species live in large colonies on intertidal sand and mud flats.

GLANS DOG WHELK *Nassarius glans*

Thinner shelled than most of the family. Tall-spired with nodulose shoulders. The early whorls bear axial ribs, but the body whorl is smooth. Sharp teeth on the lip. Cream shell with orange-brown blotches and thin brown spiral lines. Grows to 50 mm long. Intertidal sand flats. Indo-Pacific and northern Australia.

CAKE DOG WHELK *Nassarius arcularia*

Stout, solid shell with a flat, thickly callused base, nodulose shoulders and turreted spire. Cream or yellow shell, often with a central brown band and brown spots between the shoulder nodules. Grows to about 40 mm long. Active on sand of coral reef flats. Indo-Pacific and northern Australia.

Filamentous Horse Conch

Tower Latirus

Recurved Latirus

Glans Dog Whelk

Cake Dog Whelk

CORONATE DOG WHELK *Nassarius coronatus*

Rather like the Cake Dog Whelk but smaller; the callus is not so thick and does not form a flat base. May be cream, brown, pale tan or grey, often with a central brown band and white shoulder nodules; brown interior. Grows to 25 mm long. Sand on coral reef flats. Indo-Pacific and northern Australia.

IMPOVERISHED DOG WHELK *Nassarius pauperatus*

Solid and distinguished by spiral rows of nodules on the shoulders. The basal callus is moderately strong. Cream or fawn, usually with brown spiral bands. Most shells of this species are less than 15 mm long. A denizen of muddy flats in bays and estuaries. Southern Australia.

CHANNELLED DOG WHELK *Nassarius dorsatus*

The early whorls of the tall spire are axially ribbed but the remainder are smooth and glossy. The narrow basal callus terminates with a prominent spine projecting over the siphon canal. Greenish grey, cream or brown, sometimes banded. Grows to 30 mm long. Very common on muddy intertidal flats. Indo-Pacific and northern Australia.

BANDED DOG WHELK *Nassarius pyrrhus*

Heavily nodulose, tall-spired species with a narrow basal callus. White or fawn with brown spiral bands. Grows to about 20 mm long. Very common in sand in the vicinity of seagrass beds. Southern Australia from Bass Strait to Fremantle.

PIMPLED DOG WHELK *Nassarius papillosus*

Large, striking, tall-spired, heavy shell with spiral rows of prominent nodules and pointed teeth on the lip. Cream to yellowish, the nodules white and the tip of the spire dark rose. Grows to about 50 mm long. Common on sandy flats associated with coral reefs and well known throughout the Indo-Pacific and across northern Australia.

DOVE SHELLS Family Columbellidae

Dove shells are mostly small and oval to fusiform with a thickened and toothed outer lip, a smooth or plaited columella and a short anterior canal. Most species live among algae and rocks, but some crawl on sand or mud. They are either carnivorous or herbivorous. There are many minute species which may be extremely abundant in their preferred habitats. Some of the middle-sized species are among the most common shells in beach wash along southern shores.

TAPERING DOVE *Mitrella acuminata*

All the species of this genus have tall, tapering spires, this one taller than most. It has straight sides and a glossy surface. Most specimens are uniform brown with a darker brown subsutural band interrupted by white blotches but variations with axial streaks or a reticulate pattern are common. Grows to about 20 mm long. Crawls in sand among rocks and seagrass. Southern Australia.

Coronate Dog Whelk

Impoverished Dog Whelk

Banded Dog Whelk

Channelled Dog Whelk

Pimpled Dog Whelk

Tapering Dove

LINCOLN'S DOVE *Mitrella lincolnensis*
Similar to Tapering Dove but shorter and less slender. Red-brown, grey-brown, orange or yellow with irregular white patches and tent marks and often a faint overlying reticulate pattern of darker lines. Grows to 13 mm long. Crawls in sand among rocks and seagrass. Southern Australia. .

AUSTRALIAN DOVE *Mitrella australis*
Tall-spired with convex spirally grooved whorls and weakly constricted base. Fawn or pink exterior with faint white tent marks and sometimes spiral brown and white dashes; white to pink interior. Grows to about 20 mm long. Rocky shores. Eastern Australia.

LONG DOVE SHELL *Mitrella pulla*
Similar to the Australian Dove but smaller. Light brown to purple-brown exterior, usually with a faint white reticulate pattern; white anterior end. Grows to 16 mm long. Rocky shores. South-eastern Australia.

SEMICONVEX DOVE *Mitrella semiconvexa*
Similar to the Australian Dove and about the same size, but fusiform with wider and more swollen whorls. The colour is extremely variable, ranging through white, yellow, apricot, brown or mauve, sometimes bearing a pattern of brown axial or reticulate lines. Rocky shores. South-eastern and southern Australia.

STOUT DOVE *Pyrene turturina*
Very solid, almost globular shell. Like the others of this genus, the lip is thick and strongly toothed, as is the columella. White, yellow, brown or pink exterior with white tent markings; mauve lip and columella. Grows to about 15 mm long. Abundant under stones on coral and rocky reefs. Indo-Pacific and northern Australia.

LETTERED DOVE *Pyrene scripta*
Stoutly biconical with prominent shoulders and a strongly toothed lip and columella. Grows to about 18 mm long. White with zigzag axial brown and yellow lines. Very common under stones throughout the Indo-Pacific and across northern Australia.

YELLOW DOVE *Pyrene flava*
One of the larger species in the family; grows to about 30 mm long. Tall-spired with a deep suture and slight shoulders. Long, narrow aperture is weakly denticluate centrally. Colour is extremely variable but most specimens are white with thin, wavy yellow axial lines and dark brown axial streaks. Very common under stones. Indo-Pacific and across northern Australia.

Lincoln's Dove

Long Dove Shell

Stout Dove

Lettered Dove

Australian Dove

Semiconvex Dove

Yellow Dove

VOLUTES   Family VOLUTIDAE

Volutes are sand-dwelling predators, feeding on a variety of invertebrates including other molluscs. They have a long siphon and a large foot which is often colourfully patterned. The anterior canal is a wide notch. The surface of the shell is usually smooth, often highly polished, although there are axial ribs in some species; the shoulders may be nodulose or spiny. The columella has strong spiral plaits. There is no pelagic larval stage, making these snails vulnerable to local extinction. Australia is especially rich in species of this family, most of them endemic.

GRAY'S VOLUTE   *Amoria grayi*
The shells of this genus are slender with a conical spire and smooth, glossy surface. This common species is typically cream, ash or beige, sometimes pink near the lip, with a light brown band at the suture. Grows to 100 mm long. Sand across the entire width of the continental shelf. Western Australia from Geographe Bay to the Kimberley.

CAROL'S VOLUTE   *Amoria maculata*
Like Gray's Volute but yellow, cream or pale orange with four spiral rows of brown axial lines and dark brown spots in the suture. Grows to about 80 mm long. Sandy bottoms in the shallow subtidal zone; occasionally in the intertidal zone. North-eastern Australia.

WAVY VOLUTE   *Amoria undulata*
About the same size as Gray's Volute but heavier with rounded shoulders. Cream exterior with widely spaced, brown or orange-brown zigzag (undulate) axial lines and sometimes a band of brown blotches. Apricot interior. Sand in the subtidal zone, occasionally found intertidally. Eastern and south-eastern Australia.

BEAUTIFUL VOLUTE   *Cymbiola pulchra*
There are several distinct forms of this variable species, each with its own name. The typical form is elongate–oval with a tall spire and pointed shoulder nodules. Pink or orange ground colour suffused with tiny white triangles, thin brown axial lines below the suture and four red-brown, heavily spotted spiral bands. Grows to about 80 mm long. Sand in the subtidal zone, but occasionally on intertidal sand cays. Queensland. The illustrated specimen represents the form from the Capricorn Group of islands near the southern end of the Great Barrier Reef.

MITRE-SHAPED LYRIA   *Lyria mitraeformis*
Stoutly fusiform with a tall spire and prominently sculptured with broad axial folds crossed by fine spiral ribs. Cream with spiral rows of brown lines and bands of streaky brown blotches. Grows to 50 mm long. The western shells are wider than the eastern ones. Crawls in sand of the intertidal and the subtidal zones, usually in rocky areas. Southern Australia from Bass Strait to Cape Leeuwin.

Gray's Volute

Carol's Volute

Wavy Volute

Mitre-shaped Lyria

Beautiful Volute

COMMON BALER *Melo amphora*
One of four species of *Melo* in Australia, this is the largest species of the genus; grows to
500 mm long or longer. Wide, oval shell with a low spire and very wide aperture. The shoulders
are usually spiny. Yellow-brown, typically with zigzag axial lines enclosing pale tent marks and
usually two broad brown spiral bands. Glossy yellow interior. Intertidal and subtidal sand.
Northern Australia.

OLIVES AND ANCILLAS Family Olividae

Olives and ancillas are sand-dwellers in tropical and warm temperate seas. There are many
species in both northern and southern Australia. They leave trails as they crawl about hunting
their prey, which comprises a variety of other invertebrates. The foot is usually large and fleshy
and may envelope the shell. The olives have solid, highly polished shells that may be brightly
coloured. They are more or less cylindrical with a low conical spire and a long, narrow mouth.
Ancilla shells are usually more delicate with a tall spire and a small horny operculum.

RED-MOUTH OLIVE *Oliva miniacea*
Largest of the Australian olives; grows 80 mm long or longer. Cylindrical, variably coloured
shell; usually cream with wavy axial lines of orange and blue-green or blue-grey and two
purplish spiral bands. Orange interior. Sand cays on coral reefs throughout the Indo-Pacific
and Queensland but apparently not present in Western Australia.

COMMON OLIVE *Oliva oliva*
The spire of this shell is higher than most in the genus and the whorls rather more convex.
Cream or brown with profuse darker brown spots, zigzag lines and sometimes spiral bands.
Grows to about 40 mm long. Very common on sand flats in the intertidal zone. Indo-Pacific
and across northern Australia.

RINGED OLIVE *Oliva annulata*
Chunky shell with a pronounced shoulder and an incised suture. Usually cream or fawn with
a diffuse pattern of triangular brown or bluish spots and prominent spots at the suture. The
inside is yellow or orange. Grows to about 60 mm long. Sand cays and lagoons of coral reefs.
Indo-Pacific and northern Australia.

PURPLE-MOUTH OLIVE *Oliva caerulea*
Rather stout with a low spire. White or pale yellow with a pattern of diffuse zigzag axial
greenish or brown lines and faint purplish streaks at the suture. Dark violet interior. Grows
to about 60 mm long. Sand cays on coral reefs. Indo-Pacific and northern Australia.

Common Baler

Ringed Olive

Red-mouth Olive

Common Olive

Purple-mouth Olive

AUSTRALIAN OLIVE *Oliva australis*

Slender and tall-spired. Like many olives it comes in many shades but most are white with brown zigzag lines and spots and bluish flames at the suture. Grows to about 30 mm long. Sand on intertidal flats and in the shallow subtidal zone. Southern and Western Australia from Bass Strait to Broome.

ORNATE OLIVE *Oliva lignaria*

Slender and cylindrical with a low spire. Usually white or cream with brown axial lines and irregular transverse bands; some specimens are uniformly chocolate brown. Purple interior. Grows to about 50 mm long. Lives on intertidal sand flats. Indo-Pacific and across northern Australia.

CARNELIAN OLIVE *Oliva carneola*

Rather small, solid shell with a low spire. Remarkable for its white shell strongly banded with yellow or orange. Grows to about 20 mm long. A coral reef species that lives in sand on intertidal flats and in shallow lagoons. Indo-Pacific and Queensland.

MARGINED ANCILLA *Alocospira marginata*

Fusiform with a tall spire which has angular spiral lirae on the whorls. The body whorl is inflated and the lip has a prominent tooth. Cream with brown bands at the suture and anterior end. Grows to about 40 mm long. Sand in the intertidal and subtidal zones. Southern Australia.

GOLDEN ANCILLA *Ancillista velesiana*

Large, thin-shelled species with a tall spire. The body whorl is inflated and fawn with a brown band at the anterior end. The spire whorls are surrounded by a broad golden-brown band. Grows to at least 90 mm long. Not seen in the intertidal zone very often — most specimens are taken by trawlers. Eastern Australia.

MITRES AND RIBBED MITRES
Families MITRIDAE and COSTELLARIIDAE

The two families considered here are related, distinguished mainly by anatomical differences. Most species live in the shallow waters of the tropics but there are many in the temperate waters of southern Australia. The shells are usually slender with a narrow aperture and strong columellar plaits. There may be a thin periostracum but no operculum. Mitres eat worms; ribbed mitres prey on other molluscs or sea squirts.

EPISCOPAL MITRE *Mitra mitra*

The largest of the Australian mitres; grows to 180 mm long. Tall-spired with low, smooth shoulders and a toothed lip. Smooth, white exterior with spiral bands of red or orange spots; yellow interior. Common sand-dweller on coral reefs. Indo-Pacific and northern Australia.

Australian Olive

Ornate Olive

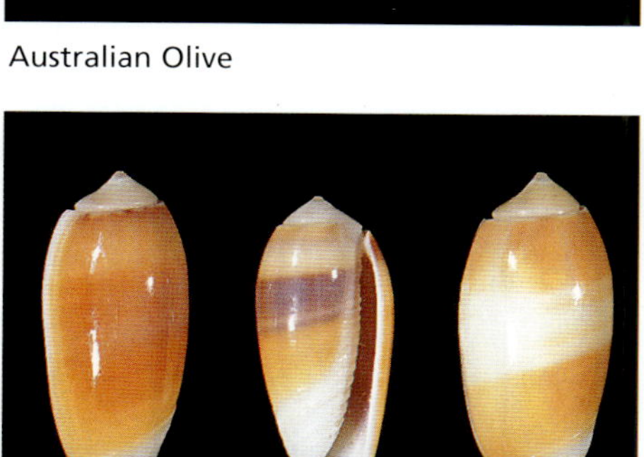

Carnelian Olive

Margined Ancilla

Golden Ancilla

Episcopal Mitre

PAPAL MITRE *Mitra papalis*

Similar to the Episcopal Mitre but with a shorter spire and coronate shoulders. Smooth white exterior with spiral rows of crimson spots; cream interior. Grows to about 160 mm long. Sand on coral reefs. Indo-Pacific and Queensland.

PONTIFICAL MITRE *Mitra stictica*

The spire of this shell is telescoped with flat-sided whorls and angular, sharply coronate shoulders. Smooth white exterior with rows of rectangular scarlet or orange blotches; yellow interior. Grows to about 80 mm long. Sand-dweller on coral reefs. Indo-Pacific and northern Australia.

CARDINAL MITRE *Mitra cardinalis*

Tumid with a tall spire. Pitted spiral grooves cover the surface. Cream with spiral rows of large squarish brown spots. Grows to about 70 mm long. Sandy lagoons of coral reefs. Indo-Pacific and Queensland.

GLABRA MITRE *Mitra glabra*

Tall, slender shell with a very high, tapering spire and weakly toothed lip. Sculptured with pitted spiral grooves near the front. Usually fawn or tan with darker axial streaks and orange-brown spiral lines. Grows to about 90 mm long. Among rocks. Southern Australia from central New South Wales to the central west coast.

PUNCTURED MITRE *Mitra puncticulata*

Solid and elongate with a ledged and coronate suture. Sculpture of spiral cords with minutely pitted interspaces. Cream or orange with dark orange-brown axial streaks tending to form two broad spiral bands. Grows to 50 mm long. Coral reefs. Indo-Pacific and northern Australia.

BUTTERFLY MITRE *Neocancilla papilio*

Lens-shaped outline with convex whorls sculptured with spiral ribs crossed by axial grooves giving the surface a nodulose texture. The lip is crenulate. White or grey ground colour with white, bluish and red-brown nodules. Orange interior. Grows to about 60 mm long. Sand-dweller on coral reefs. Indo-Pacific and northern Australia.

FILE MITRE *Cancilla filiaris*

Very beautiful fusiform shells with convex whorls and distinctive sculpture comprising prominent beaded spiral ribs and axial grooves. White to pale brown; the brown ribs stand out. Grows to 30 mm long. Sand around coral reefs. Indo-Pacific and northern Australia.

FOX RIBBED MITRE *Vexillum vulpeculum*

Slender with a tall, sharply pointed spire and narrowing anterior end. The shoulders are often nodulose and there are broad axial folds crossed by spiral striae on the whorls. Colour variable but usually orange, orange-brown or bluish with spiral bands of dark brown or black. Grows to 50 mm long. Sand-dweller on rocky flats and coral reefs. Indo-Pacific and northern Australia.

Papal Mitre

Pontifical Mitre

Cardinal Mitre

Glabra Mitre

Punctured Mitre

Butterfly Mitre

File Mitre

Fox Ribbed Mitre

CONE SHELLS Family CONIDAE

This is a very large family with at least 300 species, including many in northern Australia and several in the temperate south. Most cone shells have low spires, straight or slightly convex sides and long, narrow apertures. There is thin periostracum and, in some species, a very small operculum. Cones are predators and eat worms, molluscs or even small fish. They use a complex venom apparatus to kill their prey. The venom of the fish-eating species is particularly powerful and their sting may be fatal to humans. Live cones should not be handled.

GEOGRAPHY CONE *Conus geographus*
This large cone is easily recognised by its thin, cylindrical shell with a low, coronate spire, nodulose shoulders, convex sides and wide mouth. Bluish or white exterior with several broad bands of brown blotches and a faint reticulate pattern of brown lines enclosing small white tent-like areas. White interior. Grows to 130 mm long. A dangerous, fish-eating species responsible for many human fatalities. Shelters under stones on coral reefs. Indo-Pacific and northern Australia.

TEXTILE CONE *Conus textile*
A well-known obconical shell with rounded shoulder and a complex pattern of white tent marks outlined in brown, overlaying a ground colour of golden brown. There may be two central bands where the darker colours are concentrated. White interior. Grows to about 100 mm long. Another very dangerous species. Shelters under stones on coral and rocky reefs. Indo-Pacific and northern Australia.

LETTERED CONE *Conus litteratus*
Heavy, obconical shell with a flat spire and sharply angular shoulder. White with three indistinct yellow spiral bands, spiral rows of large squarish black to chocolate brown spots and dark wavy radial lines on the spire. Grows to 130 mm long. Common on coral reefs. Indo-Pacific and northern Australia.

MARBLED CONE *Conus marmoreus*
Large, strikingly patterned obconical shell with a coronate spire and nodulose shoulder. Black with white triangular patches; pink interior. Grows to 100 mm long. Common in the open on coral reefs among rubble and sand. Indo-Pacific and northern Australia.

FLAG CONE *Conus vexillum*
Large and wide at the shoulder with a low spire. Orange-brown with two poorly defined pale bands and a dark brown anterior end. Grows about 110 mm long. Under stones on rocky and coral reefs. Indo-Pacific and across northern Australia.

Geography Cone

Textile Cone

Marbled Cone

Lettered Cone

Flag Cone

WEASEL CONE *Conus mustelinus*
Angular shoulder and low spire. Olive green or yellow with distinct white bands at the shoulder and centre. The shoulder band has axial brown streaks and the central band is bordered by black spots. Purple interior. Grows to about 70 mm long. Under stones on rocky and coral reefs. Indo-Pacific and across northern Australia from the Kimberley to northern New South Wales.

CAPTAIN CONE *Conus capitaneus*
Similar to the preceding species but wider and more angular at the shoulder. The central spots are larger and less distinct and there are spiral rows of small spots around most of the body whorl. Grows to about 80 mm long. Under stones on rocky and coral reefs. Indo-Pacific and across northern Australia from the Kimberley to northern New South Wales.

OMARIA CONE *Conus omaria*
Slender, nearly cylindrical cone with a low spire and rounded shoulder. The surface bears an intricate reticulate pattern of red tent-shaped lines overlain by irregular red-brown blotches. Grows to about 80 mm long. Under stones in the intertidal and subtidal zones of rocky and coral reefs. Indo-Pacific and northern Australia from the Kimberley to southern Queensland.

STRIATE CONE *Conus striatus*
Large, nearly cylindrical cone that is considered to be dangerous. Its low spire has channelled sutures. Surface finely sculptured with spiral striae. Large, irregular reddish brown blotches over a cream background. Grows to about 100 mm long. Usually found in sand under stones in the intertidal and subtidal zones. Indo-Pacific and northern Australia.

SOLDIER CONE *Conus miles*
Medium-sized with a moderate spire and slightly rounded shoulder. Cream with fine yellow-brown axial lines, a light brown band around the centre and dark brown at the anterior end. Grows to 80 mm long. May be found lying among rubble on intertidal coral reefs. Indo-Pacific and across northern Australia.

CALF CONE *Conus vitulinus*
Medium-sized and heavy with a low spire and concave sides. White with two broad reddish brown spiral bands and wavy axial lines and a purple-brown anterior end. Grows to 60 mm long. Common on intertidal rock platforms and coral reefs. Indo-Pacific and northern Australia, from the Kimberley to southern Queensland.

MAGUS CONE *Conus magus*
Rather slender with nearly straight sides and a low to moderate spire which is conspicuously spirally grooved. Colour is very variable, usually white or cream with broad yellow or orange bands and spiral rows of fine brown lines and dots. Grows to 60 mm long. Under stones and corals in the intertidal and shallow subtidal zones. Indo-Pacific and northern Australia from the Kimberley to southern Queensland.

Weasel Cone

Captain Cone

Omaria Cone

Striate Cone

Soldier Cone

Magus Cone

Calf Cone

IMPERIAL CONE *Conus imperialis*

Large, solid shell with straight sides, an almost flat, coronate spire and nodulose shoulder. White or cream with broad brown bands and encircling but interrupted spiral brown lines and blue-grey anterior end. Grows to 110 mm long. May be found lying among rubble on coral reef flats. Indo-Pacific and northern Australia from the Kimberley to southern Queensland.

CAT CONE *Conus catus*

Said to be a fish-eater and possibly dangerous. Stout and solid with a low spire and beaded spiral cords. Brown with a pale central band, irregular pale bluish blotches and white dots on the cords. Grows to 50 mm long. May be found in crevices on the reef front of intertidal platforms. Indo-Pacific and northern Australia from the Kimberley to northern New South Wales.

FALSE VIRGIN CONE *Conus emaciatus*

Low spire and attenuate body whorl with slightly concave sides. Sculptured with spiral threads and axial striae. Dull yellow or orange, covered in life by a velvety periostracum; violet anterior end. Grows to 50 mm long. Crevices on intertidal coral reef platforms. Indo-Pacific and Queensland.

BLOODSTAINED CONE *Conus sanguinolentus*

How this cone came by its name is a mystery — the colouring does not suggest blood stains. Low, coronate spire and several spiral rows of small beads anteriorly. Olive green, white beads, purple anterior end and interior. Grows to about 45 mm long. Nestles into cracks and crevices on intertidal reef flats. Indo-Pacific and northern Australia.

IVORY CONE *Conus eburneus*

Stout shell with a low spire and a smooth surface. White with spiral bands of large brown spots; covered by a thin, adherent yellowish periostracum. Grows to 60 mm long. Sand-dweller in shallow sandy lagoons and intertidal sand flats associated with coral reefs. Indo-Pacific and northern Australia from the Kimberley to southern Queensland.

PERON'S CONE *Conus dorreensis*

Very unusual cone: the shell is white beneath a yellow periostracum that adheres in a wide spiral band, leaving the shell surface exposed around the shoulder and the anterior end. Rather high, coronate spire. Grows to about 40 mm long. Western Australian species found in the intertidal zone of rocky reefs from Albany to Dampier.

LIVID CONE *Conus lividus*

Coronate spire with an angular, nodulose shoulder. Olive green or sometimes yellow to orange-brown with a bluish white central band; purple anterior end and interior. Grows to 60 mm long. Cracks and crevices on intertidal rocky and coral reef platforms. Very widespread in the Indo-Pacific, across northern Australia and well south into the temperate zone of both the east and west coasts.

Imperial Cone

False Virgin Cone

Ivory Cone

Livid Cone

Cat Cone

Bloodstained Cone

Peron's Cone

YELLOW CONE *Conus flavidus*
The spire is low and the shoulder angular but not nodulose. Orange-brown with white spiral bands at the shoulder and centre; purple anterior end and violet interior. Grows to 60 mm long. Cracks and crevices on intertidal rocky and coral reef platforms. Indo-Pacific and northern Australia.

SAND-DUSTED CONE *Conus arenatus*
Stout with a low spire, slightly convex sides and rounded, slightly nodulose shoulders. White or cream profusely spotted with brown and with irregular axial and spiral bands. White interior. Grows to 60 mm long. Sand-dweller on coral reefs. Indo-Pacific and across northern Australia.

VERMICULATE CONE *Conus chaldaeus*
Solid, stoutly biconical shell with a weakly nodulose shoulder and straight sides. Blackish with narrow white or yellow spiral bands at the shoulder and centre and wavy white or yellow axial streaks. Grows to about 40 mm long. Rocky intertidal platforms. Indo-Pacific and across northern Australia from the central west coast to northern New South Wales.

ANEMONE CONE *Conus anemone*
Form and colour of this species are extremely variable. They usually have a low grooved spire and slightly convex sides. Most commonly blue-grey flecked with white, and with ragged axial lines and blotches. Some forms are pink, orange, yellow or white. Grows to as long as 80 mm, but most adult specimens are less than 50 mm long. Very common on rocky shores in the intertidal zone where they hide under stones during the day. This is the common cone of southern Australia, ranging from New South Wales to the central west coast.

PAPILLA CONE *Conus papilliferus*
Rather thin-shelled but stout and broad at the shoulder, which lacks nodules. Bluish white with blue-grey and brown blotches tending to be arranged axially; violet or brown interior. Common under stones in the intertidal and shallow subtidal zones. Eastern Australia from central Queensland to Victoria.

VICTORIA CONE *Conus victoriae*
A moderately thin-shelled species with pointed spire and convex sides. Colour variable with an off-white bluish background overlain by reticulate orange lines forming a dense tent pattern and three interrupted spiral brown bands; interior violet. Grows to 70 mm long. Common under stones in the intertidal zone. Northern Western Australia and the Northern Territory.

Yellow Cone

Sand-dusted Cone

Vermiculate Cone

Anemone Cone

Papilla Cone

Victoria Cone

AUGERS Family TEREBRIDAE

Augers have long, slender pointed shells with a small aperture and simple outer lip. The sturdy columella is plaited and the sculpture of the whorls may be axial, spiral or cancellate. Some species are quite smooth and glossy. There is a small horny operculum. These sand-dwelling predators make straight trails in the sand as they drag their shells either on or just below the surface in search of their prey, worms. There are some temperate species in southern Australia but the family is most strongly represented in the tropics.

MARLINSPIKE AUGER *Terebra maculata*
The largest species of the family; grows to 240 mm long. Massive, solid shell with convex whorls. Exterior cream with two spiral bands of dark brown rectangular patches below the suture and three fawn or tan bands on the lower part of the body whorl. Interior fawn. Sandy lagoons on coral reefs. Indo-Pacific and northern Australia.

FLY-SPOTTED AUGER *Terebra areolata*
Slender with long and slightly convex whorls and a teardrop-shaped mouth. Cream or fawn with spiral bands of large squarish brown spots, four on the body whorl. Grows to about 160 mm long. Sandy lagoons on coral reefs. Indo-Pacific and northern Australia.

SUBULATE AUGER *Terebra subulata*
Very tall, slender shell with flat-sided whorls, a deep suture and a rectangular mouth. Cream with two spiral bands of large, squarish brown spots on the spire whorls and three on the body whorl. May grow to 170 mm long. Sandy lagoons on coral reefs. Indo-Pacific and northern Australia.

CRENULATE AUGER *Terebra crenulata*
Flat-sided whorls with small rounded nodules at the suture. Aperture oval. Fawn, flesh or beige with spiral rows of brown dots and brown axial lines below the suture. Grows to about 150 mm long. Sandy lagoons on coral reefs. Indo-Pacific and northern Australia.

DIMIDIATE AUGER *Terebra dimidiata*
The whorls of this shell are slightly convex without shoulders; shallow subsutural groove. Delicate pink or pale orange with white axial and spiral lines dividing the colour into rectangular or arrow-shaped blocks. Grows to 130 mm long. Sandy lagoons on coral reefs. Indo-Pacific and northern Australia.

DUPLICATE AUGER *Duplicaria duplicata*
Members of this genus have a deep groove around the body whorl just below the suture. This species is flat-sided with broad flat axial ribs. May be white, orange, beige, bluish grey or brown, often with darker blotches and a thin pale spiral line around the base. Grows to 90 mm long. Intertidal sand flats. Central Indo-Pacific and north-western Australia.

Marlinspike Auger

Fly-spotted Auger

Subulate Auger

Crenulate Auger

Dimidiate Auger

Duplicate Auger

Sea Slugs — Opisthobranchs

This diverse group of gastropods comprises no less than nine orders and nearly 90 families. Many of them have colourful bodies but no shell. Some have fragile remnants of a shell, often embedded within the body of the animal. A few, at the more primitive end of the evolutionary scales in their respective groups, have a fully coiled shell. A small selection of the shelled species is illustrated here to represent the group.

SOUTHERN BUBBLE *Bulla quoyii*

The strong shells of this genus are swollen with a sunken spire. The aperture, which is almost as long as the shell, is narrow at the back end but wide at the front: the large, fleshy animal is able to completely withdraw within the shell. This southern species is nearly straight-sided, slightly narrower at the back. Exterior mottled brown with indistinct darker spiral bands. Grows to 50 mm long. Sand among algae and seagrasses. Southern Australia.

ROTUND BUBBLE *Bulla ampulla*

Like the Southern Bubble, and about the same length, but broader and more nearly spherical. Fawn with profuse brown flecking. Indo-Pacific and northern Australia.

PAPER BUBBLE *Hydatina physis*

Shells of this genus are tumid, very thin and fragile. The large, fleshy body is incapable of being contained within the shell. They feed on worms. The shell of this species is fawn with numerous narrow brown spiral lines; grows to about 30 mm long. The body is light brown with a mauve fringe. Among algae-covered rocks. Circum-tropical; northern Australia.

CHINESE UMBRELLA *Umbraculum sinicum*

This gross slug has an enormous, fleshy, warty foot and an uncoiled, saucer-shaped shell which is held on top of the body. The shell has a dull grey upper surface but may be polished and yellow underneath. The animals grow to as long as 80 mm, while the shell grows to about 40 mm in diameter. They feed on sponges and may be found during the day sheltering under stones. Indo-Pacific; circum-Australian.

SOLUTE BUBBLE *Akera soluta*

Stoutly cylindrical and paper thin, the shells of these slugs may be as large as 50 mm long. The translucent shell is off-white or cream. The animals are much too big to be contained within the shell. They crawl in sand among sea plants in the intertidal and shallow subtidal zones. Indo-Pacific; circum-Australian.

Southern Bubble

Southern Bubble — live animal

Rotund Bubble

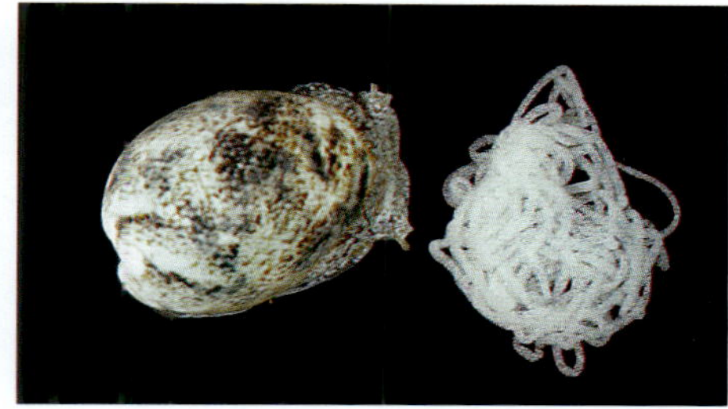

Live Rotund Bubble with egg mass

Paper Bubble

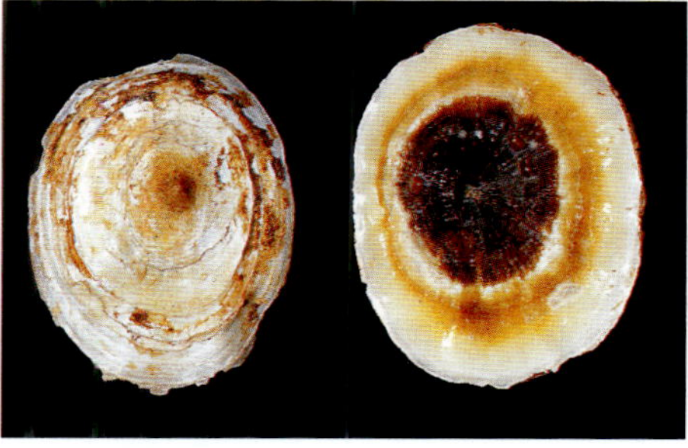

Chinese Umbrella

Solute Bubble

Solute Bubble — live animal

Air Breathers — Pulmonates

The majority of the snails and slugs that live on land have no gills but breath air by means of a pulmonary cavity or 'lung'. Several families inhabit the intertidal zone of the shore where they may be covered by sea water at high tide. Their shells range from the limpet-like siphonarians that live on rocky shores to the multi-whorled globular ear shells that crawl on estuarine mud flats. A few examples of Australian intertidal marine air breathers are illustrated here to represent the group.

ANGULATE EAR *Cassidula angulifera*
Solid, stout shell with a flat base, angular shoulders and a low, conical spire The outer lip is thickened and the elongate aperture is heavily nodulose on both sides. Variable exterior colour, usually light brown with darker brown spiral bands; white to fawn interior. Grows to more than 30 mm long. Mangrove dweller. Northern Australia.

MIDAS EAR *Ellobium aurismidae*
Solid shell with wide, slightly angular, finely nodulose shoulders. Thickened outer lip and one large spiral plait on the columella. Yellow-brown exterior, white interior. Grows to 80 mm long. High in the intertidal zone among the roots and trunks of mangrove trees. South-western Pacific and Queensland.

JUDE'S EAR *Ellobium aurisjudae*
Elongate with a tall spire, rounded shoulder and brown finely granular surface. Swollen outer lip is indistinctly toothed; callused inner lip has two strong folds. Grows to 60 mm long. Mangrove-dweller among logs and rocks high in the intertidal zone. Indo-Pacific and northern Australia.

FRAGILE AIR BREATHER *Salinator fragilis*
Fragile, globular little shell with a deep umbilicus. Brown and spirally banded. Grows to about 15 mm long. Large colonies on upper intertidal mud flats in estuaries. Southern Australia.

VAN DIEMEN'S FALSE LIMPET *Siphonaria diemenensis*
Limpet-like but with a siphonal groove on the inner surface on the right side. Light brown exterior between numerous white radial ribs; polished, pale interior, except for chestnut stripes around the edge. Grows to 20 mm long. Abundant on rocks high in the intertidal zone. Southern Australia.

ZELAND FALSE LIMPET *Siphonaria zelandica*
Like Van Diemen's False Limpet but with a very low profile, weaker radial ribs and paler colour. Grows to about 18 mm long. Abundant on rocks high in the intertidal zone. Southern Australia.

Angulate Ear

Jude's Ear

Midas Ear

Fragile Air Breather

Van Diemen's False Limpet

Zeland False Limpet

Cephalopods Class CEPHALOPODA

These highly specialised marine predators include the nautiluses, cuttlefishes, squids and octopuses. Their muscular tentacles sometimes bear suckers which they use to grasp prey, which they then bite with their hard, beak-like chitinous jaws. Some cephalopods also have poison glands associated with the mouthparts that help subdue prey.

The modern *Nautilus* and *Spirula* and the extinct ammonites possess coiled shells. Others, like the squids and most cuttlefishes, have only a remnant of the ancestral shell enclosed within the body to give it stiffening. Octopuses are shell-less, although in the breeding season female argonauts make a shell casing to contain their eggs.

Most cephalopods are nektonic animals, that is, they live their lives swimming in the ocean water column. They swim by means of muscular flaps or wings. Octopuses are the exception — they crawl on the seabed using their tentacles as 'legs'. Cephalopods are also capable of 'jet-propelled' swimming: their muscular bodies can forcibly eject water from the mantle cavity through a funnel-like siphon, producing jerky, jet-propelled movements which they may use to escape from predators. They are also able to eject a cloud of 'ink', which confuses their attacker.

The shells and shell-like artefacts of some cephalopods are spectacular finds among the flotsam cast ashore on oceanic beaches and a selection of them is illustrated here.

CHAMBERED NAUTILUS *Nautilus pompilius*

Strong shell with a mother-of-pearl inner surface, coiled planispirally and partitioned internally into a series of gas-filled flotation chambers. Manipulation of the gas density gives the animal a depth-control capacity like that of submarines. There are several species. This one is the most common and is differentiated by its closed umbilicus. White exterior with red-brown stripes. Grows to 150 mm diameter. The animals live in the deep oceans, but their shells are sometimes cast ashore on beaches. Central Indo-Pacific and northern Australia.

GIANT CUTTLEFISH *Sepia apama*

Cuttlefish are not fish at all, but squid-like creatures. They are characterised by an internal structure, made of spongy calcareous material held in a hard chitinous frame, that provides support for the body. It is called the 'cuttlebone'. After the animal has died and rotted away (or been eaten), the buoyant, white cuttlebone floats at the surface of the ocean. Cuttlebones are a very common feature on Australian beaches. The illustrated species is the largest of the Australian cuttlefish; it grows to 350 mm long. Meeting a big one under a ledge can be a frightening experience for a novice diver, but these animals are harmless. Shallow subtidal zone of the southern Australian coast.

RAM'S HORN *Spirula spirula*

Small, chambered, planispiral shells like nautilus but the coils are not in contact. May grow to 30 mm diameter. The animals live in the deep ocean worldwide. Their shells are cast ashore in thousands, especially on southern Australian beaches during wintertime.

Chambered Nautilus

Chambered Nautilus sectioned to show internal chambers

Cuttlebone of a Giant Cuttlefish

Giant Cuttlefish beneath a ledge

Ram's Horn

NODOSE PAPER NAUTILUS *Argonauta nodosa*

Paper nautiluses are octopuses, not related to the chambered nautiluses. Their fragile shells are merely temporary artefacts made by females to contain their egg masses. They live in off-shore waters. The shells are washed ashore seasonally, especially after storms in the winter months, sometimes in large numbers. This is the most common species; its shell is characterised by a double row of prominent nodules around the periphery and wavy nodulose radial ribs. The shell is white with dark brown blotches on the peripheral nodules. Grows to 130 mm diameter. Southern Ocean.

COMMON PAPER NAUTILUS *Argonauta argo*

Like the previous species in colouring but narrower and the radial ribs are simple and the periphery more sharply keeled with smaller nodules. Grows to 200 mm diameter. Worldwide.

LUNATE BLUE-RINGED OCTOPUS *Hapalochaena lunata*

There are several species of this group of small octopuses common on rocky shores all around Australia. Most of the time these octopuses are yellow-brown with darker maculations. The blue rings appear when the animal is disturbed. Like most octopuses these beautiful little creatures are shell-less but an illustration is included here because collectors are likely to encounter them frequently, and because they are so dangerous. They are very shy but must not be handled as their bite can be fatal. They often inhabit the empty shells of other molluscs. The illustrated specimen is a northern Australian species. It has a span of about 150 mm when the arms are spread.

Nodose Paper Nautilus

Common Paper Nautilus

Lunate Blue-ringed Octopus

Glossary

acuminate tapered to a point

adductor muscle muscle connecting the two valves of a bivalve (normally two)

adductor scar impression on the inner surface of a bivalve shell where an adductor muscle was attached

anterior front

anterior canal anterior notch or trough-like or tubular extension of the shell supporting the anterior siphon

aperture opening or entrance of the shell (= mouth)

apex tip of the spire

apical whorls those whorls near the apex

axial parallel or nearly so to the shell axis (= longitudinal)

axis imaginary line through the apex, about which the whorls are coiled

benthic living on the sea bed

biconic resembling two cones placed base to base

bifid divided into two parts or lobes

body whorl last and usually the largest whorl of the coiled shell

byssal gape the gap between the valves of a bivalve shell through which the byssus protrudes

byssus bundle of hairlike strands by which bivalves attach

calcareous made of lime

callus calcareous thickening

cancellate ornamented with intersecting spiral and axial ridges

cardinal tooth central hinge tooth below the umbo

chitin (adj. chitinous) a hard or flexible organic substance that forms the shells of insects and crustaceans and the periostracum and mouthparts of many molluscs

cilia minute hair-like structures on body cells aiding movement of fluids

clasper spines/fronds spines or fronds that clasp the shell to some firm object

columella pillar along the axis of a coiled shell, formed by the inner walls of the whorls and often forming a thickened inner apertural lip

columellar lip inner edge of the aperture comprising the visible part of the columella

concentric direction coinciding with growth lines — spiral in gastropods or parallel with the margin in bivalves

cord fine, round-topped spiral or axial ridge

coronate with tubercles or nodules around the shoulders of the whorls like a crown

crenulate with the edge notched

decussate with a pattern of intersecting striae

dentition in bivalves refers to the hinge teeth; in gastropods refers to nodules on the apertural lips

detrital feeders animals that feed on small particles of organic debris (detritus)

dextral coiled with a right-hand spiral, i.e. clock-wise when viewed from the apex

dorsal on the back

endemic peculiar to a region

entire uninterrupted

equilateral parts of shell valves anterior and posterior to the umbos equal in length

equivalve with two valves of equal size and convexity

excurrent water current expelled from the mantle cavity

foot the muscular ventral part of the body used for locomotion

funicle ridge spiralling into the umbilicus

fusiform spindle-shaped, swollen at the center and tapering almost equally toward the ends

gape opening remaining when shell valves are closed

genus group of genetically related species possessing certain characters in common (plural genera)

granose surface with minute grains

gregarious living in groups or clusters

growth line more or less axial line representing former position of the outer lip (in gastropods) or margin (in bivalves)

heterodont with distinct cardinal and lateral teeth

hinge the dorsal ridge of a shell valve about which the valves open and close

hinge tooth shell structure on the hinge that interlocks with hinge socket of the opposite valve

imbricate with laminae overlapping each other like roof tiles

incurrent water current drawn into the mantle cavity

inner lip edge of the gastropod aperture near the shell axis, extending from the suture to the anterior end of the columella

interspaces grooves or spaces between ribs or cords

intertidal between high- and low-tide levels

keel angular ridge

lamella thin plate-like ridge (plural lamellae)

lamellate sculpture of thin plate-like ridges

lateral tooth hinge tooth located some distance from the umbo

length in gastropods: distance from the apex to the anterior tip of the shell (sometimes also called height);

in bivalves: distance between the anterior and posterior margins
lenticular shaped like a lens (in cross-section)
ligament horny elastic structure joining the two valves of a bivalve shell that springs them apart when the adductor muscles relax
lira prominent elevated ridge or fold (plural lirae)
lirate sculpture of prominent elevated ridges or folds
lunule a depression in the dorsal margin of a bivalve shell in front of the umbo
maculate patterned with blotches i.e. with maculations
multispiral with many whorls
muscle scar impression on the inner shell surface where muscles were attached
nodulose with conspicuous nodules (= nodose)
obconical with the form of an inverted cone
obligate able to live only in a particular habitat
operculum horny or calcareous structure borne by the foot and serving for the closure of the aperture of gastropods
outer lip outer margin of the aperture of gastropods
pallial line an incised groove around the inner margin of a bivalve shell marking the attachment of the mantle muscles
pallial sinus indentation or bend in the pallial line marking the insertion of siphonal muscles
pedal muscle scar scar on the inner surface of a shell marking the insertion of the foot muscles
pelagic inhabiting the water column of the open ocean
periostracum outer layers of horny material covering the calcareous shell (= epidermis)
periphery greatest circumference of a whorl or valve
phylum primary division of the animal kingdom, e.g. Mollusca
plait spiral fold or ridge on the columellar lip
planispiral coiled in a single plane (does not form a spire)
planktonic drifting or weakly swimming in the ocean
planktotrophic feeding on other planktonic organisms
posterior rear
posterior canal notch or trough-like or tubular extension of the posterior apertural margin supporting the posterior siphon (= anal canal)
protoconch embryonic shell, present in the adult as the apical or nuclear whorls and often demarcated from the teleoconch whorls by a change of sculpture
pseudotaxodont with numerous short hinge teeth transverse to the hinge line but having a different origin to true taxodont teeth
pustulose sculpture of small pimples

pyriform pear-shaped
quadrate tending to be squarish or rectangular
radula teeth, or toothed file-like organ in the mouth of some classes of mollusc
recurved with the distal end (e.g. anterior canal) bent away from the shell axis
reflected turned outward and backward at the margin
reticulate forming a network of intersecting lines
rib round-topped elevated ridge of moderate width and prominence
serrate notched or toothed at the edge like a saw
sessile permanently attached or stationary
simple (referring to lip or spine) smooth
sinistral coiled in a left-handed spiral, i.e. anti-clockwise when viewed from the apex
sinus a deep indentation or cavity
siphon tubelike extension of the mantle for the passage of incurrent or excurrent water
spatula spoon-shaped area outlined by muscle scars on the inner surface of limpets
spatulate spoon-shaped
species group of actually or potentially inter-breeding natural populations which are reproductively isolated from other such groups
spire the coiled part of the shell consisting of all the whorls except the last
striate minutely grooved
stria narrow and shallow incised groove (plural striae)
striate marked with striae
subcircular not quite circular
subquadrate not quite square
subrhomboidal not quite rhomboidal
subspecies a geographically defined group of populations comprising individuals which possess characteristics distinguishing them from other such subdivisions of the species
subsutural below the suture
subtidal below low-tide level
sulcate grooved or furrowed
supratidal above high-tide level
suture continuous line on the shell surface where successive whorls adjoin
taxodont with numerous short hinge teeth more or less transverse to the hinge line
terminal umbos umbos at the front end
tessellate with a colour pattern consisting of regular patches
trigonal triangular in outline
truncate cut off at the end, blunt
tubercle protuberance or knob

tumid swollen

turbinate turban-shaped

turreted with the spire whorls like a succession of turrets emerging one above the other

umbilicus cavity or hollow around which the axis of gastropod shells forms when the inner walls of successive whorls do not meet

umbo apex of a bivalve shell above the hinge

valve separate part of a molluscan shell

varicose with varices (= varixed)

varix elevated ridge formed by a thickened and reflected former outer lip (plural varices)

veliger shelled larval stage of molluscs

ventral underside

ventricose swollen

whorl any complete coil of a spiral shell

Further Reading

This list contains books on marine shells and shell collecting in Australia. However, many of the tropical Australian shells are widely distributed in the Indian and Pacific Oceans and books published in the countries of those regions are also relevant to the study of Australian shells. Countless taxonomic studies too numerous to list here deal with particular groups of molluscs irrespective of their geographic distribution and may also contain relevant material. For more comprehensive lists of references see the publications by Lamprell & Whitehead (1992) and Lamprell & Healy (1998) for the bivalves and by Wilson (1993 & 1994) for the gastropods, and the technical works dealing with all the Australian molluscs contained within the two volumes edited by Beesley et al. (1998).

Allen, J. (1959). *Australian Shells*. Georgian House, Melbourne.

Beesley, P. L., Ross, G. J. B. & Wells, A. (eds) (1998). *Mollusca: the Southern Synthesis. Fauna of Australia*. 5. CSIRO Publishing: Melbourne, Part A xvi pp. 563, Part B viii pp. 565–1234.

Coleman, N. (1981). *What Shell is That?*, second edition. Ure Smith Press, Sydney.

Coleman, N. (1976). *Shell Collecting in Australia*. A. H. & A. W. Reed, Sydney.

Coleman, N. (1981). *Shells Alive*. Rigby, Sydney.

Cotton, B. C. & Godfrey, F. K. (1940). *The Molluscs of South Australia, Part 2, Scaphopoda, Cephalopoda, Aplacophora and Crepipoda*. Government Printer, Adelaide.

Cotton, B. C. (1959). *South Australian Mollusca — Archaeogastropoda*. Government Printer, Adelaide.

Cotton, B. C. (1961). *South Australian Mollusca — Pelecypoda*. Government Printer, Adelaide.

Hinton, A. (1972). *Shells of New Guinea and the Central Indo-Pacific*. Robert Brown and Associates, Port Moresby.

Hinton, A. (1978). *Guide to Australian Shells*. Robert Brown and Associates, Port Moresby.

Jansen, P. (1995). *Seashells of Central New South Wales*. Privately published.

Jansen, P. (1996). *Common Seashells of Coastal North Queensland*. Privately published.

Lamprell, K. & Whitehead, T. (1992). *Bivalves of Australia*. Volume 1. Crawford House, Bathurst.

Lamprell, K. & Healy, J. (1998). *Bivalves of Australia*. Volume 2. Backhuys, Leiden.

Macpherson, J. H. & Gabriel, C. J. (1962). *Marine Molluscs of Victoria*. Melbourne University Press. (Victorian National Museum Handbook No. 2.)

May, W. L. (1958). *An Illustrated Index of Tasmanian Shells* (revised edition by J. H. Macpherson) Government Printer, Hobart.

McMichael, D. F. (1960). *Shells of the Australian Sea Shore*. Jacaranda Press, Brisbane.

Rippingale, O. H. & McMichael, D. F. (1961). *Queensland and Great Barrier Reef Shells*. Jacaranda Press, Brisbane.

Short, J. W. & Potter, D. G. (1987). *Shells of Queensland and the Great Barrier Reef — marine gastropods*. Robert Brown and Associates, Bathurst.

Wells, F. E. & Bryce, C. W. (1985). *Seashells of Western Australia*. Western Australian Museum, Perth (revised 1990).

Wells, F.E. & Bryce, C.W. (1993). *Seaslugs of Western Australia*. Western Australian Museum, Perth.

Wilson, B. R. & Gillett, K. (1971). *Australian Shells*. A. H. & A. W. Reed, Sydney (revised 1974).

Wilson, B. R. & Gillett, K. (1979). *Field Guide to Australian Shells*. A. H. & A. W. Reed, Sydney (reprinted 1985, 1988).

Wilson, B.R. (1993). *Australian Marine Shells*. Volume 1. Odyssey Publishing, Perth.

Wilson, B.R. (1994). *Australian Marine Shells*. Volume 2. Odyssey Publishing, Perth.

Index